GCSE 9-1

geography

AQA

Revision Guide

Rebecca Tudor

Tim Bayliss

Catherine Hurst

Series editor

Bob Digby

OXFORD
UNIVERSITY PRESS

OXFORD
UNIVERSITY PRESS

Great Clarendon Street, Oxford, OX2 6DP, United Kingdom

Oxford University Press is a department of the University of Oxford.
It furthers the University's objective of excellence in research, scholarship, and
education by publishing worldwide. Oxford is a registered trade mark of Oxford
University Press in the UK and in certain other countries

British Library Cataloguing in Publication Data

Data available

ISBN 978-019-842346-1

Kindle edition ISBN 978-019-842347-8

10 9 8 7

Printed in India by Multivista Global Pvt. Ltd.

Acknowledgements

The publisher and authors would like to thank the following for permission to
use photographs and other copyright material:

Cover: watchara/Shutterstock; **p8:** Chris Wildt/Cartoonstock; **p11:** Bob Digby;
p12: Stocktrek Images, Inc./Alamy Stock Photo; **p21(l):** MARTIN BERNETTI/
Getty Images; **p21(r):** Manish Swarup/REX/Shutterstock; **p22:** Tim Bayliss;
p25: Harvepino/Shutterstock; **p27:** imagegallery2/Alamy Stock Photo; **p28:**
Neil Cooper/Alamy Stock Photo; **p29:** Apex News and Pictures; **p30:** SWNS/
Alamy Stock Photo; **p31:** Crown coyright (2017) Ordinance Survey; **p32:** David
Moir/REUTERS; **p33:** NASA; **p36:** EDF Energy; **p37(t):** Jack Sullivan/Alamy
Stock Photo; **p37(b):** Pix/Alamy Stock Photo; **p40:** Land and Water Services ;
p42: Travel Ink/Getty Images; **p43(t):** Mint Images/Frans Lanting/Getty Images;
p43(b): Bazuki Muhammad/Reuters; **p44:** © Eye Ubiquitous/Alamy; **p45:** pyzata/
Shutterstock; **p46:** Brian Bailey/Getty Images; **p48(l):** Franck METOIS/Alamy
Stock Photo; **p48(r):** Tibor Bognar/Getty Images; **p49(t):** Bartek Wrzesniowski/
Alamy Stock Photo; **p49(b):** Soltan Frédéric/Getty Images; **p50:** Avalon/Photoshot
License/Alamy Stock Photo; **p51:** Eye Ubiquitous/REX/Shutterstock; **p52:** © Eye
Ubiquitous/Alamy Stock Photo; **p53(t):** Frans Lanting/Getty Images; **p53(b):** Tyler
Olson/Shutterstock; **p54(t):** Renato Granieri/Alamy Stock Photo; **p54(m):** Dmitry
Chulov/Shutterstock; **p54(b):** © Rolf Adlercreutz/Alamy Stock Photo; **p55:** © ASK
Images/Alamy Stock Photo; **p56(t):** © Robert Harding Picture Library Ltd/Alamy
Stock Photo; **p56(b):** MarcAndreLeTourneux/Shutterstock; **p64(b):** Mike Charles/
Shutterstock; **p64(t):** dbphots/Alamy Stock Photo; **p65(l):** Crown coyright (2017)
Ordinance Survey; **p65(r):** Angie Sharp/Alamy Stock Photo; **p67(l):** Mick House/
Alamy Stock Photo; **p67(r):** Westend61/Getty Images; **p68:** © Environment
Agency 2017; **p72:** Paul Heinrich/Alamy Stock Photo; **p74(t):** Washington
Imaging/Alamy Stock Photo; **p74(b):** Crown coyright (2017) Ordinance Survey;
p76: David Angel/Alamy Stock Photo; **p81:** David Robertson/Alamy Stock Photo;
p82: Crown coyright (2017) Ordinance Survey; **p83(tl):** Simon Ross; **p83(tr):**
Jeff Morgan 08/Alamy Stock Photo; **p83(bl):** Yorkman/Shutterstock; **p83(br):**
©Natural Retreats; **p84:** Terry Abraham; **p88:** America LLC/Alamy Stock Photo;
p89: imageBROKER/Alamy Stock Photo; **p92(t):** Peter Treanor/Alamy Stock
Photo; **p92(m):** antonio di paola/Alamy Stock Photo; **p92(b):** imageBROKER/
Alamy Stock Photo; **p93:** Peter Tsai Photography/Alamy Stock Photo; **p94(t):**
lazyllama/Shutterstock; **p94(b):** Mario Tama/Getty Images; **p97(l):** urbanbuzz/
Istockphoto; **p97(r):** Doug Houghton/Alamy Stock Photo; **p98(l):** Pictorial
Press Ltd/Alamy Stock Photo; **p98(r):** Russell Binns/Alamy Stock Photo; **p99(r):**
Scott Hortop Travel/Alamy Stock Photo; **p100(t):** pjhpix/Shutterstock; **p100(b):**
Jane Tregelles/Alamy Stock Photo; **p102(t):** Jeff Morgan 04/Alamy Stock Photo;
p102(b): Courtesy of Leese and Nagle; **p103(b):** Bennett Dean /Eye Ubiquitous/
Getty Images; **p104(l) :** Bristol City Council; **p104(r):** Crown coyright (2017)
Ordinance Survey; **p105(t):** PLEIADES © CNES 2016, Distribution Airbus DS;
p105(b): Populous Arena Team; **p106:** Hemis/Alamy Stock Photo; **p107(l):**
LOOK Die Bildagentur der Fotografen GmbH/Alamy Stock Photo; **p107(r):** Rolf
Disch Solar Architecture, Germany; **p107(b):** Daniel Schoenen/LOOK-foto/
Getty Images; **p108(t):** allOver images/Alamy Stock Photo; **p108(b):** PETER
PARKS/Getty Images; **p114:** © Jenny Matthews/Alamy Stock Photo; **p115:**
Kjell Nilsson-Maki, cartoonstock.com; **p116:** Nick Turner/Alamy Stock Photo;
p117: © Danita Delimont/Alamy Stock Photo; **p118(t):** Design Pics Inc/REX
Shutterstock/; **p119(t):** © flowerphotos/Alamy Stock Photo; **p119(b):** FAIRTRADE;
p120: Majority World/REX Shutterstock; **p121(t):** Andrew Park/Shutterstock;
p121(b): © robertharding/Alamy Stock Photo; **p124:** epa/Shutterstock; **p125:** Eye
Ubiquitous/Alamy Stock Photo; **p126:** Unilever Nigeria; **p128(l):** Sipa Press/REX/
Shutterstock; **p128(tr):** Peeter Viisimaa/Getty Images; **p128(br):** Nick Turner/
Alamy Stock Photo; **p129:** © jordi clave garsot/Alamy Stock Photo; **p131(t):**
Corbis; **p131(b):** robertharding/Alamy Stock Photo; **p132:** www.morecobalt.
co.uk; **p133(l):** © Construction Photography/Alamy Stock Photo; **p133(r):** Image
courtesy of AGGREGATE INDUSTRIES UK LIMITED; **p134:** Shutterstock; **p136:**
Peel Ports Group Ltd; **p138:** Paul Lovelace/REX Shutterstock; **p139:** Cameron
Spencer/Getty Images; **p146(t):** Abbie Trayler-Smith/Panos; **p146(b):** Images of
Africa Photobank/Alamy Stock Photo; **p147:** Maya Pedal; **p149:** Jim West/Alamy
Stock Photo; **p163(t):** Newzulu/Alamy Stock Photo; **p163(b):** Bob Digby; **p165:**
Andy Slater; **p167(l):** Hannah Peters/Getty Images; **p167(r):** Nigel Spiers/Alamy
Stock Photo; **p168:** NASA; **p176:** Shutterstock; **p177:** Bob Digby;

Artwork by Aptara Inc., Mike Connor, Barking Dog Art, Simon Tegg, and Q2A
Media Services Inc.

Every effort has been made to contact copyright holders of material reproduced
in this book. Any omissions will be rectified in subsequent printings if notice is
given to the publisher.

Contents

Contents

Unit 2 – Challenges in the human environment

Contents

Guided answers are available on the Oxford Secondary Geography website:
www.oxfordsecondary.co.uk/aqa_gcse_geog

Please note this revision guide has not been written or approved by AQA. The answers and commentaries provided represent one interpretation only and other solutions may be appropriate.

Introduction: Helping you succeed

If you want to be successful in your exams, then you need to revise all you've learnt in your GCSE course! That can seem daunting – but it's why this book has been written. It contains key revision points that you need to prepare for exams for the AQA GCSE 9–1 Geography specification.

Your revision guide!

This book is designed to help you revise for your three AQA GCSE Geography exam papers.

Each unit is split into sections. Each section has an introduction page which contains an outline of:

• the three exam papers you'll be taking
• the key ideas and content that form the specification.

Each page of your revision guide has the following features:

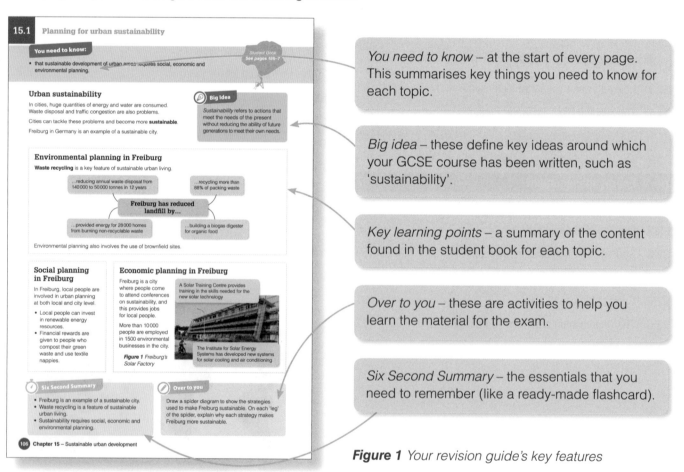

You need to know – at the start of every page. This summarises key things you need to know for each topic.

Big idea – these define key ideas around which your GCSE course has been written, such as 'sustainability'.

Key learning points – a summary of the content found in the student book for each topic.

Over to you – these are activities to help you learn the material for the exam.

Six Second Summary – the essentials that you need to remember (like a ready-made flashcard).

Figure 1 *Your revision guide's key features*

Your revision!

Each topic (1.1, 1.2, etc.) in this revision guide exactly matches the content for each topic in your GCSE Geography AQA student book. Key content in each double page in the student book is summarised in a single page in this revision guide.

The AQA GCSE 9–1 Geography specification has three units. Each unit contains topics. Each unit is assessed by an exam (Paper 1, 2, 3) with sections for sets of different topics, as follows:

Unit 1 Living with the physical environment

This is assessed by Paper 1 in the exam. It has three sections, each with different topics.

- **Section A** *The challenge of natural hazards* includes topics on Tectonic hazards, Weather hazards, and Climate change.

- **Section B** *The living world* includes topics on Ecosystems, Tropical rainforests, Hot deserts and Cold environments. You have to know about Ecosystems and Tropical rainforests, and **either** Hot deserts **or** Cold environments.

- **Section C** *Physical landscapes in the UK* includes topics on UK physical landscapes, Coastal landscapes, River landscapes, and Glacial landscapes. You have to know about UK physical landscapes and **any two** of Coastal, River or Glacial landscapes.

In addition, there'll be questions on geographical skills (e.g. how to interpret statistics, maps, diagrams or photos) in every topic.

Complete this for Unit 1!

The three *Living world* topics I've studied are Ecosystems, Tropical rainforests and

_____.

The two Physical landscapes in the UK I've studied are

_____

_____ and

_____

_____.

Unit 2 Challenges in the human environment

This is assessed by Paper 2 in the exam. It also has three sections, each with different topics.

- **Section A** *Urban issues and challenges* includes topics on a case study of a major city in **either** a low-income country (LIC) **or** a newly emerging economy (NEE), **and** a case study of a major UK city.

- **Section B** *The changing economic world* includes topics on a case study of **either** an LIC **or** an NEE, **and** Economic futures in the UK.

- **Section C** *The challenge of resource management* includes topics on Resource management, Food, Water and Energy. You have to know about Resource management and **one** from Food **or** Water **or** Energy.

Like Paper 1, there'll be questions on geographical skills (e.g. how to interpret statistics, maps, diagrams or photos) in every topic.

Complete this for Unit 2!

The major city I've studied in **either** an LIC **or** an NEE is

_____

The major city I've studied in the UK is

_____.

The LIC or NEE that I've studied is

_____.

I've studied Resource management and

_____

Unit 3 Geographical applications

This is assessed by Paper 3 in the exam. It has two sections:

- **Section A** *Issue evaluation* will be based on a pre-release booklet that you will receive a few weeks before the exam. You will study this booklet in class time.

- **Section B** *Fieldwork* will be about your two days' fieldwork (one day physical, one day human), including questions on links between physical and human geography.

Complete this for Unit 3!

My *Issue evaluation* enquiry topic is

_____.

The two days of fieldwork that I've done are

Physical geography:

_____

Human geography:

_____.

However you look at it, revision can be boring! But, whatever revision you do should be **active**. This page will help you develop useful ways of revising.

Revising in groups

Working as a group is always better than alone. Try these ideas out.

Form a study group with friends. Join two or three friends and fix times when you'll go through key topics. Do timed questions together, then mark them. Make lists of things you don't understand to ask your teacher.

Working together at home. Message, Facetime or Skype friends and test each other. Go through questions together.

Know your key words. Make lists of key words that you need to know.

Test each other. Make flash cards of key words and have revision competitions.

"We prefer to call this test 'multiple choice,' not 'multiple quess.'"

Figure 2 *The dangers of not revising!*

Revising in class

You'll have lessons to revise topics that you're not clear about. Use the time well!

Get to know question styles. Know command words, practise timed answers and plan longer extended answers.

Get to know how exam questions are marked. Know which questions are point marked (one correct point = 1 mark) and those marked using levels 1, 2 and 3. Know what qualities are required for the highest levels. Look at sample mark schemes, available on AQA's website and GCSE 9–1 Geography AQA Exam Practice (ISBN 9780198423485).

Look at past answers. Some exam boards publish model answers, or have marking exercises as part of their training for teachers. Go through these, so that you know how examiners mark.

Extra lessons. Make lists of questions about things you don't understand about past exam questions, then see your teacher to go through them.

Ask your teacher for revision help. Your teacher can give you questions on particular topics you are less confident about.

Revising alone

At some stage, you'll have to revise alone! Don't sit in front of the TV trying to read notes that you're not sure about! Try these variations.

Act on weaknesses. Make a checklist of things you need to know, such as those in the section introductions in this book

Work on past exam papers. The more papers you try, the more familiar you'll be with examiners' style.

Watch video clips from YouTube, BBC Bitesize or other websites. Allow no more than 15 minutes, which is as long as most people can concentrate fully.

Making a revision timetable

Planning your revision

Revision doesn't just happen – for it to go well, it needs to be well planned! Here are some handy hints to help you plan.

Two to three months before exams begin (so, probably during March), draw up a revision timetable. A plan for a school day is shown in Figure **3**.

Draw up a plan for your school holidays. Look at Figure **4** for an example plan.

Divide up the time between the subjects and topics that you need to revise. Figures **3** and **4** assume you study ten subjects.

In Figure **4**, every revision day during the school holidays has three time slots – morning, afternoon, and evening. Use **two** of these on each day; give yourself some free time. Both timetables have one day and at least two evenings completely free.

Split the three time slots in Figure **4** into three 50-minute chunks (see page 10).

Figure 3 *Revision timetable for term time (yellow blocks indicate free time)*

Time	Mon	Tues	Wed	Thurs	Fri	Sat	Sun
9 a.m.–4 p.m.	Normal school day					Free day	Normal school homework
							Subjects 9 and 10
4–6 p.m.	Normal school homework						Free afternoon
6.30–8 p.m.	Subjects 1 and 2	Subjects 3 and 4	Subjects 5 and 6	Subjects 7 and 8	Free evening		Free evening

Figure 4 *Revision timetable for school holidays (yellow blocks indicate free time)*

Time	Mon	Tues	Wed	Thurs	Fri	Sat	Sun
9 a.m.–12 p.m.	Subject 1 (3 × 50-min sessions)	Free morning	Subject 5 (3 × 50-min sessions)	Subject 7 (3 × 50-min sessions)	Free day	Subject 9 (3 × 50-min sessions)	Free morning
2–5 p.m.	Free afternoon	Subject 3 (3 × 50-min sessions)	Subject 6 (3 × 50-min sessions)	Subject 8 (3 × 50-min sessions)		Free afternoon	Subject 10 (3 × 50-min sessions)
6–9 p.m.	Subject 2 (3 × 50-min sessions)	Subject 4 (3 × 50-min sessions)	Free evening	Free evening		Free evening	Flexible

Three stages of revision

The worst thing you can do is to stare at a book of notes! Be **active** about revising and the time will fly past. Split your time up into 50-minute chunks and use Figure **5** to help you plan what to do.

Figure 5 *Three stages of revision – a traffic light approach.*

Stage 1 Mending the gaps in your knowledge and understanding	Select a topic you're unclear about, where there are gaps (perhaps you missed work) or which you find more difficult. • Read through the topic you want to revise using your own work and also the student book. • Make a list of key words. • Write a definition of each key word. Use your notes and the student book to help.
Stage 2A Revising uncertain topics The first 50 minutes	Select a topic you're becoming clearer about. • Read through the topic you want to revise using your own work and also the student book. • As you read, copy sub-headings on a sheet of file paper, and leave gaps ready for some notes. • In the gaps, write out questions about the topic as though you were an examiner.
Stage 2B Becoming more confident	Go back to your headings, questions and spaces. • Now fill in answers to questions as far as you can, without looking at notes or your student book. Don't worry about the ones you can't answer. • When you finish you can see at a glance what you know and what you don't know. • For any gaps, go back and look at your notes and student book. Then do the questions again. • Read through any past exam questions or practice questions that you've been given and make a list of the topics you think you know well.
Before Stage 3	List your strong and weak topics based upon marks in tests, and gaps in your understanding from the stages above. Be honest about your strengths and weaknesses!
Stage 3 Revising topics fully	Focus on the topics you now know well in your revision timetable. • Try the practice questions from your student book. • Either self-assess using the mark schemes, or peer mark questions done by friends using the mark scheme. • Hand in your practice question answers to your teacher for marking. • Review.

Over to you

1   Make a copy of a revision timetable for:

  • term time
  • school holiday time.

2   Go through each topic that you need to learn (see page 7), and rate it red, amber or green using the table above to guide you.

Know your key words!

In your GCSE course, you've been learning to *think* like a geographer – about the processes affecting the Earth's surface, and how people affect the natural environment. Now, you must learn to *write* like a geographer, which means knowing key geographical words.

Key words help you to:

- understand a question (e.g. explaining how a process occurs)
- identify features in diagrams or photos, like Figure **6**.
- use key words in your answers. Answers that use key words earn more marks than those that don't.

- Beach
- Larger sediment particles
- Finer beach material
- Evidence of slumping
- Cliff face

Figure 6 *A coastline in Cornwall – which of these key terms can you recognise?*

Many words used in geography are used in day-to-day life like 'population', or 'beaches'. But you only ever meet some key words in certain topics (e.g. precipitation). Many are key to the subject – like the words in Figure **7**.

Do you know these key words?

In the left-hand column are definitions of terms you should know. Write them in the right-hand column 1 to 5, then check the answers on page 12.

Figure 7 *Which key words are these?*

Description	Geographical term
The rock type of an area, or the study of rocks	1
People coming to live in a country from overseas	2
Gases which warm the atmosphere	3
Specialist service and technical industries	4
The total value of goods and services in a country in a year	5

Mmmm ... mnemonics!

Learning key words and processes can be hard. Even something like spit formation has ten steps to remember (see below). However, **mnemonics** can help – it's a way of developing a system to help your memory.

How mnemonics work

1   Try this way of learning processes that form a coastal spit.

   a) Winds blow at an **angle**

   b) Waves break – **swash**

   c) Sand moved **up** the beach

   d) Water moves down the beach – **backwash**

   e) Sand moves down the beach – forming a **zigzag**

   f) This is called **longshore** drift

   g) Sand reaches an **estuary**

   h) Beach forms a **spit**

   i) River currents form a **hook**

   j) **Mudflats** form behind the spit

2   Look at the words in bold – then take the first letter – **A**ngle, **S**wash, **U**p, **B**ackwash, **Z**igzag, **L**ongshore, **E**stuary, **S**pit, **H**ook, and **M**udflats.

3   Write the letters in a list, then create a sentence from the first letters to help you remember – it can be as daft as you like!

4   Learn the sentence!

Try making mnemonics for:

a) the erosion process on rivers (11.2), coasts (10.3) or glaciers (12.1)

b) key words in any **one** other topic that you've revised.

Revising examples and case studies

Hazards is one of the most popular topics in geography. Hazard events, such as the Haiti earthquake of 2010 (Figure **8**), are among the many examples and case studies that you need to revise for the exam.

- You can use this approach for any hazard – for example, flood, volcanic eruption, storm.
- You can adapt this approach to revise **any** example or case study.

Figure 8 *The Haiti earthquake in 2010*

1

Build up a factfile

Revising case studies can be hard because there's a lot of detail. Start by building up a basic **factfile**, such as:

- Location – where is it? Can you locate it in an atlas?
- What kind of hazard was it?
- When did it occur? Date and time?
- Was it a single event or one of several?
- Describe what happened. For example, for an earthquake – what date, what time of day, how strong was it, where was the epicentre?
- What were its causes? Is it on a destructive plate margin?

2

Know its impacts

The impacts, or effects, of many hazard events can be significant. What were its short-, medium- and long-term impacts? Be clear that you know what this means.

- Short-term means within the first month or so.
- Medium-term means within six months.
- Long-term means anything longer than six months, maybe even years.

Next, classify these into economic, social or environmental impacts.

- Economic impacts (related to money, e.g. jobs, businesses, trade, costs).
- Social impacts (about people, health, and housing)
- Environmental impacts (about changes to the surrounding landscape).

You can now list these impacts and classify them, using the grid in Figure **9**.

Impact	Immediate/short-term	Medium-term	Long-term
Economic			
Social			
Environmental			

Figure 9 *A table for classifying the impacts of a hazard event such as a volcanic eruption or an earthquake*

 Over to you

Think how you could adapt a) the factfile b) the table of impacts to other topics. Consider these possibilities:

1 A study of a country (an LIC or an NEE). What sort of factfile would you produce?

2 A study of a megacity in an LIC or an NEE? How would you build up notes on reasons for its growth, or problems that it is trying to solve?

3 A study of a city in the UK (e.g. Bristol)? How would you build up notes on ways in which it is trying to become more sustainable?

Answers to Figure 7 Page 11

1 Geology
2 Immigration (or immigrants)
3 Greenhouse gases
4 Quaternary industries
5 GDP

Improving your exam answers

How to get the best marks possible in your exams? To reach your best standard of writing on longer answers, follow the advice below.

1

Plan your answer

It helps to organise your thoughts if you plan your answer. You don't need a long plan – just something that takes you 30 seconds to jot down. Some people plan using a spider diagram, others just make a short list. It helps you to get the order of the answer right and makes sure you don't forget what to write. One example of a plan is given in Figure **10**.

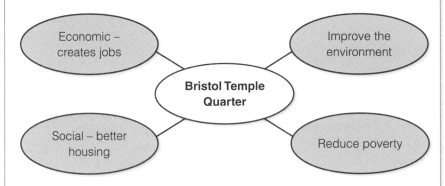

Figure 10 *An example of a plan to a 6-mark question:* 'Discuss ways in which one major UK city you have studied has been regenerated. (9 marks)'
Hint: *This refers to pages 182–5 in the student book. You could add more details to this.*

 Over to you

In the box to the left, draw a plan like the one in Figure **10** for the question:

'Discuss ways in which one city you have studied is trying to become more sustainable. (9 marks)'

Hint: you could use pages 188–9 in the student book to help you.

Improving your exam answers

Use key words

Below are two exam answers where candidates A and B were supposed to use key words.

- Candidate answer A gained only 2 marks out of 4
- Candidate answer B gained only 3 marks out of 6

In each case, the candidate knew the general idea, but there were no key words.

Over to you

1 Below are two candidate answers from an exam. To the right of each answer are key geographical words. Replace each word in bold in the answer with one of the key geographical words.

2 Write the improved answer in the space to get full marks.

Candidate answer A

Describe two ways in which human activity leads to climate change. (4 marks)

People are using more and more cars, and the gases that are **given off** go up into the **air** and **damage** the air we breathe and **make it warmer**. By **cutting down trees**, less oxygen is produced and there is more **of other gases** in the air making the **weather** warmer.

_____
_____
_____
_____
_____
_____

Key geographical words to improve the answer:
atmosphere
carbon dioxide
climate
emitted
deforestation
increase the greenhouse effect
pollute

Candidate answer B

Explain how human actions can increase the risk of flooding. (6 marks)

Human actions can increase flood risk. First, **cutting down trees** so there is **grass for cattle to feed on** and fuel means that there are no more trees to **break the fall of rain** so it gets to the soil straight away. More rain gets to the river by **flowing over the land** and the river **fills up** very quickly and **overflows**. Tree roots also bind the soil together and, if **the trees are cut down**, the soil **wears away**. This soil then gets into the river and gets **carried along** and **dropped** on the **bottom of the river** which raises it and reduces **how much water the river can hold**.

_____
_____
_____
_____
_____
_____
_____
_____
_____

Key geographical words to improve the answer:
capacity
deforestation
deforested
deposited
erodes
floods
grazing
intercept
reaches capacity
river bed
surface runoff
transported

Top tips for exam success

You'll often hear students say 'Good luck' to each other as they enter the exam room. If you have done certain things, you won't need luck! Exam success comes from following a few rules. Students who perform well almost always follow these rules:

They revise. Lack of revision always catches up with you. GCSEs are tougher now than in previous years and it's important to know your stuff!

They know which **topics** will be in each exam – for example, which exam tests physical or human geography, and which of their topics, such as hot deserts or cold environments, are on which paper.

They look at the **marks**, and know what sort of questions carry the highest marks.

They **practise** answers, often under timed conditions, for example, allowing 4 minutes for a 4-mark question.

They get **timing** right. GCSE 9–1 Geography AQA Exam Practice have marks that are similar to the length of the exam. For example, Paper 1 is 88 marks in 90 minutes. So, if a question has 6 marks, take 6 minutes. Don't get carried away on one question.

They **answer all the questions** that should be answered, and leave no blanks. Even if unsure, they write something. Leaving a 6-mark answer blank could mean giving up a whole grade.

They write in **full sentences**. Single words or phrases are fine for questions of 1–2 marks, but 4-mark answers written out in 'bullet points' rarely score well.

They learn **specific details** about case studies or examples. They take time to learn one or two statistics, names of places, and schemes. They don't just say 'in Africa'! Use specific place knowledge – you need this to earn the highest marks.

Get to know the **mark scheme** and how questions are marked. Longer answers are marked in **levels** based on quality. Examiners use three words to describe answers: Level 1, basic (no named places or examples); Level 2, general (some key points); and Level 3 detailed (facts, data, and examples). Make sure you are in Level 3!

Finally, make sure you have a **timetable** that tells you exactly which day and times every exam is on! Check, and double-check it!

Section A
The challenge of natural hazards

Your exam

Section A The challenge of natural hazards makes up part of Paper 1: Living with the physical environment.

Paper 1 is a one-and-a-half hour written exam and makes up 35 per cent of your GCSE. The whole paper carries 88 marks (including 3 marks for SPaG) – questions on Section A will carry 33 marks.

You need to study all the topics in Section A – in your final exam you will have to answer questions on all of them.

Tick these boxes to build a record of your revision

Your revision checklist

Spec key idea	Theme	1	2	3
1 Natural hazards				
Natural hazards pose major risks to people and property	1.1 What are natural hazards?			
2 Tectonic hazards				
Earthquakes and volcanic eruptions are the result of physical processes	2.1 Distribution of earthquakes and volcanoes			
	2.2 Physical processes at plate margins			
The effects of, and responses to, tectonic hazards vary between areas of contrasting levels of wealth	2.3 The effects of earthquakes – Chile and Nepal			
	2.4 Responses to earthquakes – Chile and Nepal			
Management can reduce the effects of tectonic hazards	2.5 Living with the risk from tectonic hazards			
	2.6 Reducing the risk from tectonic hazards			
3 Weather hazards				
Global atmospheric circulation helps to determine patterns of weather and climate	3.1 Global atmospheric circulation			
Tropical storms (hurricanes, cyclones, typhoons) develop as a result of particular physical conditions	3.2 Where and how are tropical storms formed?			
	3.3 The structure and features of tropical storms			
Tropical storms have significant effects on people and the environment	3.4 Typhoon Haiyan – a tropical storm			
	3.5 Reducing the effects of tropical storms			
The UK is affected by a number of weather hazards	3.6 Weather hazards in the UK			
Extreme weather events in the UK have impacts on human activity	3.7 The Somerset Level Floods, 2014 (1)			
	3.8 The Somerset Level Floods, 2014 (2)			
	3.9 Extreme weather in the UK			
4 Climate change				
Climate change is the result of natural and human factors, and has a range of effects	4.1 What is the evidence for climate change?			
	4.2 What are the natural causes of climate change?			
	4.3 What are the human causes of climate change?			
Managing climate change involves both mitigation (reducing causes) and adaptation (responding to change)	4.4 Managing the impacts of climate change (1)			
	4.5 Managing the impacts of climate change (2)			

Student Book
See pages 8–9

What is a natural hazard?

Natural **hazards** are sudden, severe events which make the natural environment difficult to manage. They disrupt human life, and have huge **economic** and **social impacts**. They fall into three main groups (Figure 1).

These events are not hazards if they occur in unpopulated regions. But where they cause high levels of death, injury, damage or disruption then they become *disasters*.

Most natural disasters are linked to four hazards:

- floods – the most frequent and deadly, damaging most property
- tropical storms – the next most frequent, almost as dangerous
- earthquakes
- droughts.

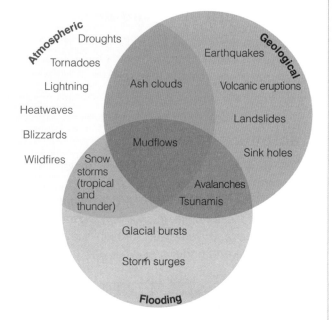

Figure 1 *Different types of natural hazard*

What is 'hazard risk'?

Hazard **risk** means the chance of being affected by a natural hazard. For example, those living near the sea at risk of flooding caused by tropical storms or tsunami. People live in risky areas because they:

- accept the risk, after weighing up advantages and disadvantages
- have little choice of where to live.

Big Idea

Just as not all hazards are disasters, not all hazards have to be feared (see 2.5).

What factors affect risk?

As populations grow, more people are exposed to natural hazards. Four factors increase the risk.

- *Urbanisation* – densely-populated urban areas concentrate those at risk.
- *Poverty* – expense of housing leads to building on risky ground.
- *Farming* – the attraction of nutrient-rich floodplains puts people at risk.
- *Climate change* – global warming raises sea levels and generates more extreme weather.

Six Second Summary

- Natural hazards are environmental events threatening people.
- Natural disasters occur where death and destruction result.
- As populations grow, so does hazard risk.

Over to you

The number of earthquakes or volcanic eruptions is not changing – so why are more people at risk from natural hazards?

Student Book
See pages 10–11

You need to know:

- the global pattern of earthquakes and volcanoes
- plate tectonics (crust, plates and plate margins).

The pattern of earthquakes

An **earthquake** is a sudden, violent period of ground-shaking. Most occur at the margins of slowly-moving tectonic plates. Friction and sticking between plates create enormous pressures and stresses which build to breaking point.

Tectonic plates

- The Earth's crust is split into seven major and several minor tectonic plates.
- There are two types of crust – dense, thin *oceanic* crust and less dense, thicker *continental* crust.
- Plates move, driven by convection currents within the mantle and under gravity.

- Plates separate at **constructive** margins forming new crust, causing volcanic eruptions.
- Plates collide at **destructive** plate margins causing subduction, earthquakes, volcanic eruptions and fold mountains.
- Plates slide by each other at **conservative** margins, causing earthquakes.

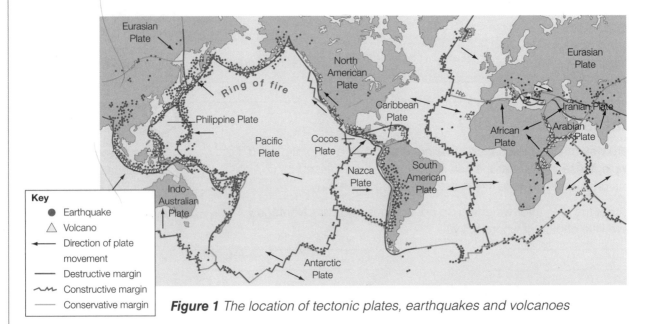

Figure 1 *The location of tectonic plates, earthquakes and volcanoes*

Distribution of volcanoes

Volcanoes are large, often cone-shaped landforms, formed over long periods by several eruptions. They are fed by molten rock (magma) deep within the Earth's mantle. Like earthquakes, most volcanoes occur in belts along plate margins (e.g. the 'Pacific Ring of Fire' and the Mid-Atlantic Ridge). But some occur at *hot spots* where the crust is thin and magma breaks through the surface (e.g. Hawaiian Islands).

 Six Second Summary

- Plate movement and tectonic activity at plate margins cause earthquakes and volcanoes.

 Over to you

- Name **three** plate margins where earthquakes occur.
- Give **one** reason why more earthquakes occur than volcanic eruptions.
- Explain why earthquakes and volcanoes occur at plate margins.

Physical processes at plate margins

2.2

You need to know:

* the physical processes at constructive, destructive and conservative plate margins.

*Student Book
See pages 12–13*

What happens at tectonic plate margins?

Constructive margin

The two plates move apart and magma forces its way to the surface. As it breaks the crust it causes mild earthquakes. The magma is very hot and fluid allowing the lava to flow a long way before cooling. This results in typically broad and flat *shield volcanoes* (e.g. Mid-Atlantic Ridge).

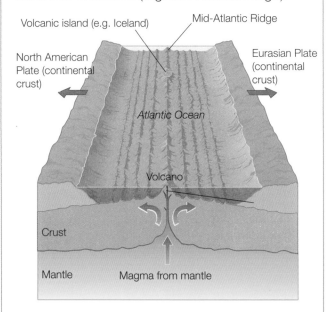

Destructive margin

Two plates move towards each other (e.g. west coast of South America, Figure **2**). Here, the dense oceanic plate is *subducted* beneath the less dense continental plate. Friction causes strong earthquakes. The sinking oceanic plate creates sticky, gas-rich magma. This results in steep-sided *composite volcanoes* which erupt violently.

Where two continental plates meet there is no subduction, so no magma to form volcanoes. The crust crumples and lifts to form fold mountains (e.g. Himalayas). Powerful earthquakes can be triggered.

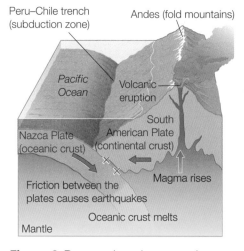

Conservative (transform) margin

Two plates move past each other at different rates. Friction between the plates build stresses and trigger earthquakes when they slip. There are no volcanoes because there is no magma.

 Six Second Summary

* Plates *separate* at constructive margins causing mild earthquakes and volcanic eruptions.
* Plates *collide* at destructive margins causing strong earthquakes and violent volcanic eruptions.
* Plates *slide by* at conservative margins causing powerful earthquakes.

Over to you

Summarise in a table the physical processes that happen at **each** type of plate margin.

The effects of earthquakes

You need to know:

- the primary and secondary effects of earthquakes in contrasting countries – Chile in 2010 and Nepal in 2015.

Example

The earthquakes in Chile and Nepal

Earthquakes can have devastating social effects requiring **immediate** and **long-term responses**:

- **Primary effects** (caused by ground shaking) destroy buildings and infrastructure, and kill and injure.
- **Secondary effects** (resulting from the shaking) include fires and landslides.
- Responses include emergency care, support and longer-term reconstruction.

Figure 1 *Location of Chile and Nepal*

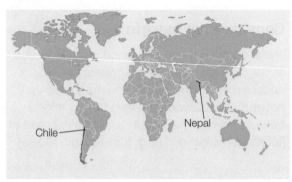

Figure 2 *Comparison of Chile and Nepal earthquakes*

	Chile, February 2010	**Nepal, April 2015**
Wealth (GDP) and quality of life (HDI) indicators	• GDP 38th out of 193 countries • HDI 41st out of 187 countries	• GDP 109th out of 193 countries • HDI 145th out of 187 countries
Cause	Nazca Plate subduction beneath the South American Plate, just off the coast of central Chile	Indo-Australian Plate colliding with the Eurasian Plate
Size	Magnitude 8.8, shallow focus (35 km)	Magnitude 7.9, very shallow focus (15 km)
Primary effects	• 500 killed, 12 000 injured and 800 000 people affected • Significant destruction of buildings and infrastructure • Power, water and communications cut • US$30 billion damage	• 9000 killed, 20 000 injured and 8 000 000 people affected • Widespread destruction of buildings and infrastructure • Power, water, sanitation and communications cut • US$5 billion damage
Secondary effects	• Communities cut off by landslides • Coastal towns devastated by tsunami • Chemical plant fire near Santiago forced evacuations	• Communities cut off by landslides and avalanches • Avalanches on Mount Everest killed at least 19 people • Flooding caused by (landslide) blocked rivers

 Six Second Summary

- Both earthquakes had primary and secondary effects.
- Both earthquakes had devastating effects on people's lives and activities.
- Contrasts in wealth and development affected the impacts.

Over to you

Study information in the table.

- Name **three** indicators (figures) which show that Nepal is poorer than Chile.
- Highlight **five** main similarities and **five** main differences between the two earthquakes.

Student Book
See pages 16–17

Example

Comparing immediate and long-term responses

Earthquakes in Chile are quite common. Both government and local communities are prepared, experienced and wealthy enough for rapid and effective response.

Earthquakes in Nepal are not uncommon. Scientists are familiar with the risks, but poverty prevents widespread adoption of new building regulations or effective preparation.

Figure 1 *Immediate and long-term responses to Chile and Nepal earthquakes*

	Chile, February 2010	**Nepal, April 2015**
Immediate responses Search, rescue and short-term aid keeping survivors alive by providing medical care, food, water and shelter	• Swift and effective response by emergency services • Key roads repaired within 24 hours • Most power and water restored within 10 days • US$60 million national appeal built 30 000 emergency wooden shelters	• Overseas aid included widely active NGOs (e.g. Oxfam) • Aid included helicopters for search, rescue and supply drops in remote areas, such as on Mount Everest • 300 000 people migrated from Kathmandu for shelter and support from family and friends
Long-term responses Rebuilding and reconstruction, to restore normal life and reduce future risk	• Strong economy reduced need for foreign aid • Government reconstruction plan to help 200 000 households • Full recovery within four years likely	• Roads repaired, landslides cleared and flood lakes drained • International conference to seek technical and financial support • Indian border blockade in 2015 caused crippling fuel, medicine and construction material shortage

Figure 2 *Temporary wooden shelters for the homeless in Chile*

Rubble to be shifted

Rescue dogs

Listening for survivors

Local knowledge

Lifting equipment

Weak buildings – danger of collapse

Video cameras to see inside collapsed buildings

Figure 3 *Searching for survivors in Kathmandu, Nepal*

 Six Second Summary

* Chile was prepared, experienced and wealthy enough for a rapid and effective response.
* Nepal's response was hindered by poverty, and it depended on overseas countries to provide aid.

 Over to you

Practise describing **three** immediate and **three** long-term responses to **each** of the disasters in Chile and Nepal.

Student Book
See pages 18–19

You need to know:

- why people continue to live in areas at risk from earthquakes and volcanoes
- how tectonic activity in Iceland brings huge benefits.

Living at risk from tectonic hazards

The majority of **tectonic hazards** occur at plate margins, some of which run through densely populated regions such as Japan, parts of China and southern Europe.

Poor people have no choice – money, food and family are seen as more important

Plate margins often coincide with favourable areas for settlement and trade, e.g. flat, coastal areas

Earthquakes and volcanic eruptions are rare, so not seen as a great threat

Some people have no experience or knowledge of the risks

Why people live at risk from tectonic hazards

Earthquake-resistant building designs reduce risk

Volcanoes can bring benefits such as fertile soils, rich mineral deposits and hot water

Effective monitoring of volcanoes and tsunami waves allow evacuation warnings to be given

Earthquake fault lines can allow water to reach the surface – important in arid regions

Figure 1 *Why people live at risk from tectonic hazards*

Living on a plate margin: Iceland

Iceland straddles the Mid-Atlantic Ridge with volcanic eruptions on average every five years. But awareness and monitoring reduces the threat to low risk. Indeed, the tectonic activity brings huge benefits, such as the Hellisheidi combined heat and power (CHP) plant which serves Reykjavik.

Figure 2 *The Hellisheidi CHP plant is the largest geothermal power plant in the world*

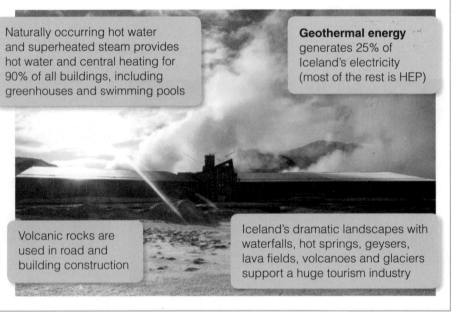

Naturally occurring hot water and superheated steam provides hot water and central heating for 90% of all buildings, including greenhouses and swimming pools

Geothermal energy generates 25% of Iceland's electricity (most of the rest is HEP)

Volcanic rocks are used in road and building construction

Iceland's dramatic landscapes with waterfalls, hot springs, geysers, lava fields, volcanoes and glaciers support a huge tourism industry

Six Second Summary

- Plate margins run through densely populated regions such as Japan and southern Europe.
- Effective monitoring, prediction and protection reduce the risks.
- Iceland benefits from tectonic activity with geothermal power and tourism.

Over to you

List **three** key reasons to explain **each** of the following:

- why people live at risk from tectonic hazards
- how the people of Iceland benefit from living on a plate margin.

You need to know:

- how risks from tectonic hazards can be reduced by monitoring, prediction, protection and planning.

Student Book
See pages 20–1

How can risks from tectonic hazards be reduced?

There are four main **management strategies** for reducing the risk from tectonic hazards.

1

Monitoring

Volcanoes

All active volcanoes are now monitored using hi-tech scientific equipment including:

- *Remote sensing* – satellites detect heat increases
- *Seismicity* – seismographs record microquakes
- *Ground deformation* – laser beams measure changes in the shape of the ground.

Earthquakes

Earthquakes generally occur without warning. A number of events can occur before an earthquake strikes, but these are never certain enough to issue warnings. These events include:

- microquakes before the main tremor
- bulging of the ground
- raised groundwater levels.

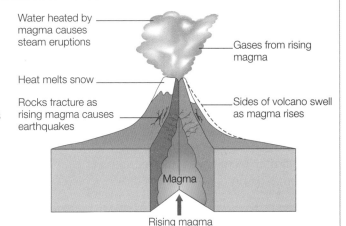

Figure 1 *Warning signs of a volcanic eruption*

2

Prediction

Volcanoes

Monitoring is now allowing accurate prediction and effective evacuation (e.g. Eyjafjallajökull, Iceland, 2010).

Earthquakes

Accurate predictions are impossible due to a lack of clear warning signs. But historical records can help determine probability and so help in planning for them (e.g. Istanbul's risk on the North Anatolian Fault, Turkey).

3

Protection

Volcanoes

Little can be done to protect property, but earth embankments and explosives have been used to successfully divert lava flows (e.g. Mount Etna, Italy).

Earthquakes

Earthquake drills help keep people alert and prepared. But earthquake-resistant construction is the best way to reduce risk.

4

Planning

Volcanoes

Risk assessment and hazard mapping to identify areas to practise evacuation or restrict building.

Earthquakes

Risk assessment and hazard mapping to identify areas to protect buildings and infrastructure.

Six Second Summary

- The risk from tectonic hazards can be reduced by monitoring, prediction, protection and planning.
- Buildings can be constructed to be earthquake-resistant.

 Over to you

Write **three** questions about the material on this page (with answers) to test a friend.

- how global atmospheric circulation works to affect global weather and climate
- examples of the effects in the UK, deserts and at the Equator.

Student Book
See pages 22–3

What is global atmospheric circulation?

The atmosphere is the air above our heads (Figure **1**) on which we depend for life.

Atmospheric circulation involves a number of interconnected circular air movements called cells (Figure **2**).

- Sinking air creates high pressure, and rising air creates low pressure.
- Surface winds move from high to low pressure, transferring heat and moisture from one area to another.
- These winds curve due to the Earth's rotation and change seasonally as the tilt and rotation of the Earth causes relative changes in the position of the overhead sun.

Add a *WOW!* factor

In the exam, use an annotated sketch or diagram if it makes your answer clearer. But remember that the marks are in your annotation, not in the quality of the drawing.

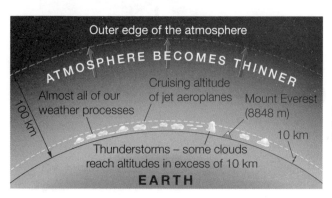

Figure 1 *The atmosphere*

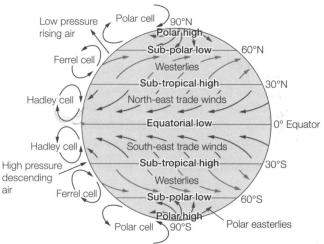

Figure 2 *Global atmospheric circulation*

How does global circulation affect the world's weather?

Global atmospheric circulation drives the world's weather:

- *Cloudy and wet in the UK* because 60° north is close to where cold polar air from the north meets warm subtropical air from the south. These surface winds from the south-west usually bring warm and wet weather, because rising air cools and condenses forming clouds and rain.

- *Hot and dry in the desert* because most deserts are found at about 30° north and south where sinking air means high pressure, little rain, hot daytime temperatures and very cold nights.
- *Hot and sweaty at the Equator* because low pressure marks where the sun is directly overhead. Hot, humid air rises, cools and condenses, causing heavy rain – hence the tropical rainforests.

 Six Second Summary

- Atmospheric circulation involves interconnected cells of air.
- Atmospheric circulation drives the world's weather.

Over to you

Practise drawing **two** annotated sketches – one to explain tropical rainforests, and another to explain deserts.

Student Book
See pages 24–5

You need to know:

- what a tropical storm is
- where tropical storms form
- how tropical storms form.

What is a tropical storm?

Tropical storms are huge storms called hurricanes, cyclones and typhoons in different parts of the world (Figures **1** and **2**). They form 5–15° north and south of the Equator, in summer and autumn, where:

- ocean temperatures are highest (above 27 °C)
- the spinning (Coriolis) effect of the Earth's rotation is very high
- intense heat and humidity makes the air unstable.

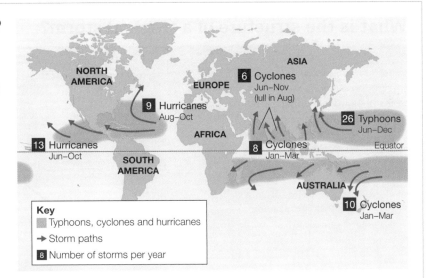

Figure 1 *The distribution of tropical storms*

Figure 2 *Satellite image of Hurricane Sandy off the coast of Florida, USA, 2012*

How do tropical storms form?

It is not certain how tropical storms are formed, but this sequence is always involved:

- Rising air draws evaporated water vapour up from the ocean surface which cools and condenses to form towering thunderstorm clouds.
- The condensing releases heat which powers the storm and draws up more water vapour.
- Multiple thunderstorms join to form a giant rotating storm.
- Coriolis forces spin the storm at over 120 km/h (75 mph) creating a vast cloud spiral with a central, calm eye of rapidly descending air.
- Prevailing winds drift the storm over the ocean surface like a spinning top, gathering strength as it picks up more and more heat energy.
- On reaching land the energy supply (evaporated water) is cut off and the storm will weaken.

Six Second Summary

- Tropical storms form 5–15° north and south of the Equator, in summer and autumn, when ocean temperatures are highest.
- They are triggered by the upward movement of evaporated air and moisture.
- They gather strength drifting over the ocean surface, but weaken over land.

Over to you

Make sure you can locate and name tropical storms associated with different parts of the world.

Student Book
See pages 26–7

You need to know:

- the structure and features of tropical storms
- how climate change might affect tropical storms in the future.

What is the structure of a tropical storm?

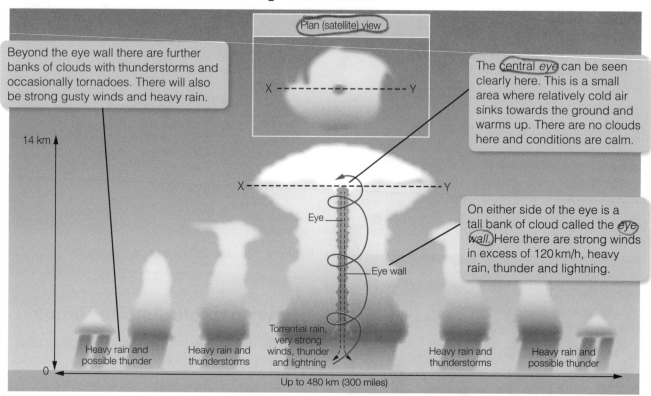

Plan (satellite) view

Beyond the eye wall there are further banks of clouds with thunderstorms and occasionally tornadoes. There will also be strong gusty winds and heavy rain.

The central eye can be seen clearly here. This is a small area where relatively cold air sinks towards the ground and warms up. There are no clouds here and conditions are calm.

14 km

Eye

Eye wall

On either side of the eye is a tall bank of cloud called the eye wall. Here there are strong winds in excess of 120 km/h, heavy rain, thunder and lightning.

Heavy rain and possible thunder

Heavy rain and thunderstorms

Torrential rain, very strong winds, thunder and lightning

Heavy rain and thunderstorms

Heavy rain and possible thunder

Up to 480 km (300 miles)

Figure 1 *The structure of a tropical storm*

Will climate change affect tropical storms?

There is strong scientific evidence of global warming and that this may be impacting on natural systems including the distribution, frequency and intensity of tropical storms:

- Over the last few decades sea surface temperatures in the Tropics have increased by 0.25–0.5 °C.
- In the future, tropical storms *may* extend into the South Atlantic and parts of the sub-tropics.
- In the future, tropical storms *may* become more powerful (as measured on the Saffir-Simpson scale).
- In the North Atlantic, six of the ten most active years since 1950 *have* happened since the 1990s.
- In the North Atlantic, hurricane intensity *has* risen in the last 20 years.

But currently there is no clear evidence that the numbers or intensities of storms are increasing. More data will be needed over a longer period of time.

Six Second Summary

- Tropical storms are the most destructive storms on Earth.
- There is strong scientific evidence of global warming, including sea surface temperatures.
- Currently there is no clear evidence that the numbers or intensities of storms are increasing – more data is needed.

Over to you

Practise drawing a simplified, labelled sketch of Figure **1**. Important labels would include 'cloud spiral', 'eye' and 'eye wall'. The horizontal and vertical scales are also crucial.

You need to know:

- the primary and secondary effects of Typhoon Haiyan
- the immediate and long-term responses to Typhoon Haiyan.

Student Book
See pages 28–9

Example

'Super' Typhoon Haiyan, November 2013

- One of the strongest Category 5 storms ever recorded (Figure **1**).
- Very low air pressure caused 5 m storm surge swept on shore by winds up to 275 km/h (170 mph).
- Coastal devastation included 90% of Tacloban destroyed by storm surge.

N
0
200
km
Gusts of up to 269 km/h
Manila
Mindoro
PHILIPPINES
Samar
Tacloban
Panay
Cebu
Leyte
Cebu
Bohol
60–120 km/h
120+ km/h
120+ km/h
60–120 km/h
Key
Affected people
More than 500 000
100 000–499 000
10 000–99 999
1000–9999
100–999
No data

Figure 1 *The track of Typhoon Haiyan*

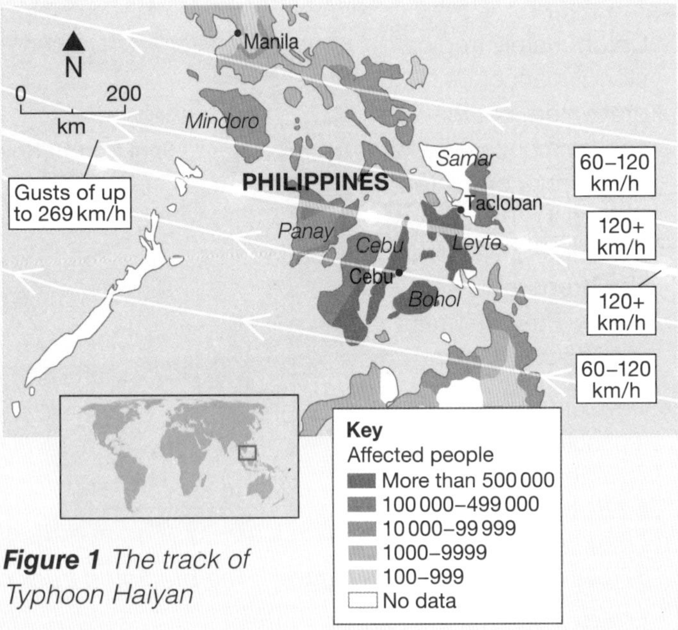

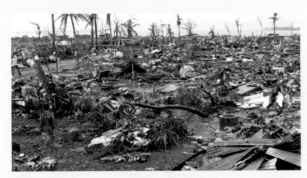

Figure 2 *Destruction in Tacloban*

Primary effects (impacts of strong winds, heavy rain and storm surge)	**Secondary effects** (longer-term impacts resulting from primary effects)
• 6300 killed – most in storm surge • Over 600 000 displaced • 40 000 homes destroyed or damaged • Wind damage to buildings, power lines and crops • Over 400 mm of rain caused widespread flooding	• 14 million affected including 6 million jobs lost • Flooding caused landslides – blocking roads and restricting access for aid workers • Shortages of power, water, food and shelter, leading to outbreaks of disease • Infrastructure including schools destroyed • Looting and violence in Tacloban
Immediate responses	**Long-term responses**
• Rapid overseas aid included NGOs • US helicopters assisted search and rescue, and delivery of aid • Field hospitals helped injured • Over 1200 evacuation centres set up	• UN and international financial aid, supplies and medical support • Rebuilding of infrastructure • Rice farming and fishing quickly re-established • Homes rebuilt in safer areas • More cyclone shelters built

Figure 3 *Effects and responses of Typhoon Haiyan*

Six Second Summary

- Typhoon Haiyan was one of the strongest storms ever recorded, destroying farms, homes, buildings, infrastructure and jobs.
- UN, international governments and NGOs responded with immediate aid and longer-term help.

Over to you

- Study Figure **3**. Learn **three** bullet points in **each** of the four segments of the table.

You need to know:

- how the effects of tropical storms can be reduced by monitoring, prediction, protection and planning.

*Student Book
See pages 30–1*

Monitoring, prediction and protection

Unfortunately tropical storms cannot be prevented, but they can be monitored and their tracks predicted (Figure 1). This allows warnings to be issued and preparations made. 'Preparedness' is all about planning.

Developments in technology, including satellite tracking, allow prediction maps to be prepared and warnings issued. For example:

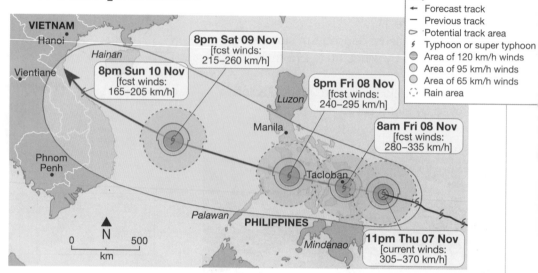

Figure 1 *The predicted track of Typhoon Haiyan*

Key
← Forecast track
— Previous track
⬭ Potential track area
⚡ Typhoon or super typhoon
◉ Area of 120 km/h winds
◎ Area of 95 km/h winds
○ Area of 65 km/h winds
⬚ Rain area

8pm Sat 09 Nov
[fcst winds: 215–260 km/h]

8pm Sun 10 Nov
[fcst winds: 165–205 km/h]

8pm Fri 08 Nov
[fcst winds: 240–295 km/h]

8am Fri 08 Nov
[fcst winds: 280–335 km/h]

11pm Thu 07 Nov
[current winds: 305–370 km/h]

- The government of the Philippines sending out Tropical Cyclone Warning Signals graded on the severity of winds and time frame expected.
- The National Hurricane Center in Miami, Florida, USA using a simpler two-scale warning system of Hurricane Watch (advised) and Hurricane Warning (expected).

Protection

Methods of protection usually involve anticipation in design – everything from reinforced walls, roofs and window shutters, to storm drains and sea walls. Cyclone shelters in Bangladesh are used as community centres, schools or medical centres for most of the time (Figure **2**).

Constructed of strong concrete

Bicycles used to give warnings to remote communities

Stairs to take people to the safety of the first floor

Shutters over windows

Built on stilts in case of floods

Built on raised ground

Figure 2 *Cyclone shelter in Bangladesh*

Planning

'Preparedness' is all about contingency planning for the inevitable. It is unrealistic to stop tens of millions of people living in coastal areas at risk from tropical storms, but they can be made safer. It mostly means education and media campaigns raising individual and community awareness in order that people understand the dangers, and are able to respond.

Six Second Summary

- Tropical storms can be monitored, their tracks predicted, and warnings issued.
- Buildings can be protected and cyclone shelters built.
- Contingency planning raises awareness allowing people to respond.

Over to you

Summarise what can, and what cannot be done in terms of monitoring, predicting and protecting from tropical cyclones.

You need to know:

Student Book
See pages 32–3

- how the UK is affected by thunderstorms, prolonged rainfall, drought and extreme heat, heavy snow and extreme cold, and strong winds
- why extreme weather occurs.

What are the UK's weather hazards?

Weather describes the day-to-day conditions of the atmosphere – temperature, rain and so on. Climate describes the average weather over a 30-year period.

Despite its moderate climate, the UK does experience weather hazards – occasional **extreme weather** events linked with its 'roundabout' location (Figure **1**) at the meeting point of different types of weather from different directions:

- *Thunderstorms* follow hot weather bringing lightning and torrential rainfall linked with 'flash' flooding (Figure **2**).

- *Prolonged rainfall* over a long period leads to river floods, such as the very wet winter of 2013/14 causing **flooding** across much of southern England.

- *Drought and extreme heat* cause rivers to dry up and reservoirs to run dangerously low. The record-breaking 2003 heatwave over much of Europe, including the UK, killed over 20 000 people – mostly young children, the frail and elderly.

- *Heavy snow and extreme cold* are less common nowadays, but can cause great hardship to people in the north of the UK. December 2010 was the coldest in a century!

- *Strong winds*, such as in February 2014, cause disruption to power supplies, damage from fallen trees and coastal battering from large waves.

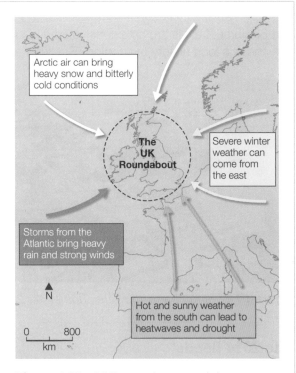

Arctic air can bring heavy snow and bitterly cold conditions

The UK Roundabout

Severe winter weather can come from the east

Storms from the Atlantic bring heavy rain and strong winds

N

Hot and sunny weather from the south can lead to heatwaves and drought

0   800
    km

Figure 1 *The UK's weather roundabout*

Figure 2 *Boscastle, Cornwall, flash flood, August 2004*

Example

You need to know:

- the causes, impacts and responses to flooding on the Somerset Levels in 2014.

Student Book
See pages 34–5

Where are the Somerset Levels?

The Somerset Levels are an extensive area of low-lying land in south-west England. They have a long history of flooding (Figures **1** and **2**).

Figure 1 *Flooding near Bridgwater, Somerset, 2014*

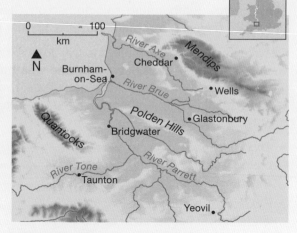

Figure 2 *The Somerset Levels*

The 2014 floods

Figure 3 *Causes, impacts and responses of the Somerset Levels floods, 2014*

Causes	Immediate responses
• A sequence of south-westerly depressions brought record rainfall in January and February. • High tides and storm surges swept water up the rivers from the Bristol Channel preventing normal flow. • Rivers, clogged with sediment, had not been dredged for 20 years.	• Huge media interest was generated. • Cut-off villagers used boats for transport. • Community groups and volunteers gave invaluable support.
Social, economic and environmental impacts	**Longer-term responses**
• Over 600 houses flooded and 16 farms evacuated. • Villages cut off – disrupting work, schools and shopping. • Estimated £10 million damage. • 14 000 ha of farmland flooded and 1000 livestock evacuated. • Power supply, roads and railway cut off. • Floodwaters contaminated with sewage, oil and chemicals. • Massive debris clearance required.	• £20 million Flood Action Plan launched by Somerset County Council and Environment Agency to reduce future risk. • 8 km of Rivers Tone and Parrett dredged. • Road levels raised in lowest dips. • Vulnerable communities will have flood defences. • River banks raised and strengthened, and more pumping stations built. • Possible tidal barrage at Bridgwater by 2024.

 Six Second Summary

- Exceptional flooding caused by record rainfall, high tides and storm surges.
- Severe impacts included villages isolated and farmland flooded.
- A Flood Action Plan will reduce future risk.

Over to you

Learn **three** of **each** of the causes, and the social, economic and environmental impacts of the 2014 floods.

You need to know:

Student Book
See pages 36–7

* how to use a 1:25 000 map to find out about flooding on the Somerset Levels in 2014.

Example

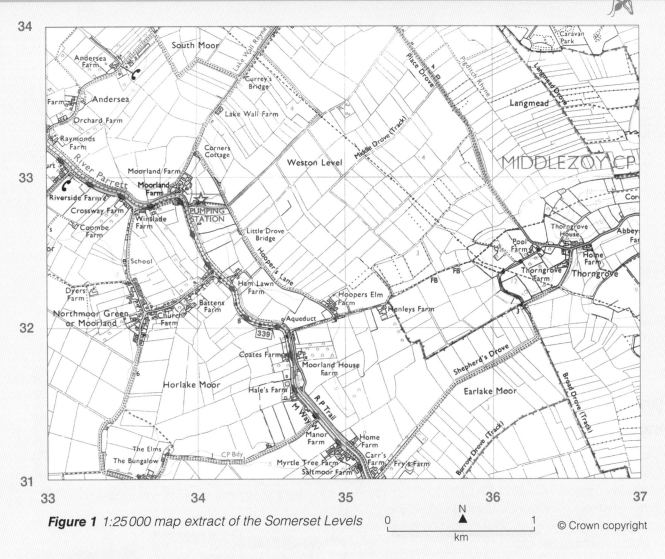

Figure 1 *1:25 000 map extract of the Somerset Levels*   0 | N ▲ | 1   © Crown copyright
km

Over to you

Practising map and photograph interpretation

1 Use the map (Figure 1) to answer the following questions.

(a) What is the evidence from the map that this area is very flat and low-lying?

(b) Why does the area have so many drainage ditches?

(c) Comment on the likelihood of flooding in Thorngrove (GR 365325).

(d) What is the evidence that most of this area is farmland?

2 Study the photograph on the opposite page (page 37 in the student book). The photograph shows part of the flooded village of Moorland (named as Northmoor Green or Moorland on the map extract at GR 337321). The church is at the road junction in the centre of the village.

(a) In what direction is the photograph looking?

(b) What is the name of the farm at the top left of the photograph?

(c) What has been done to stop this property from flooding?

(d) Describe the extent of flooding in the photograph.

You need to know:

- if the UK's weather is becoming more extreme.

Student Book
See pages 38–9

Evidence that UK weather is becoming more extreme?

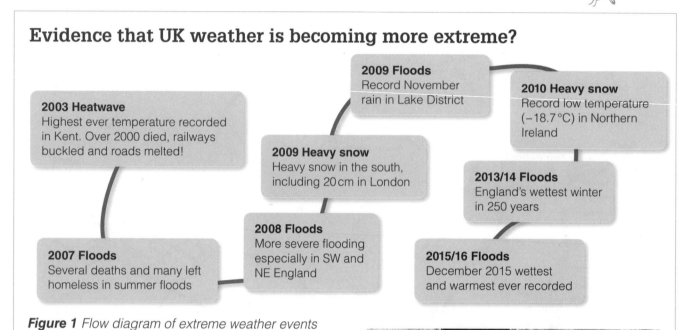

2003 Heatwave
Highest ever temperature recorded in Kent. Over 2000 died, railways buckled and roads melted!

2009 Floods
Record November rain in Lake District

2010 Heavy snow
Record low temperature (−18.7 °C) in Northern Ireland

2009 Heavy snow
Heavy snow in the south, including 20 cm in London

2013/14 Floods
England's wettest winter in 250 years

2007 Floods
Several deaths and many left homeless in summer floods

2008 Floods
More severe flooding especially in SW and NE England

2015/16 Floods
December 2015 wettest and warmest ever recorded

Figure 1 Flow diagram of extreme weather events

Why might extreme weather events be on the increase?

Recent extreme weather events have also occurred elsewhere in the world – such as the severe droughts in western USA (2014). Whilst no single weather event can be blamed on climate change, trends over many years could be linked to global warming, which:

- leads to more energy in the atmosphere, which could lead to more intense storms
- possibly affects atmospheric circulation, bringing floods to normally dry areas and heatwaves to normally cooler areas.

Figure 2 Snow causes traffic chaos in 2010

⏱ **Six Second Summary**

- The UK has experienced an increase in the number of extreme weather events in recent years.
- Scientists believe that the global increase in extreme weather events may be linked to climate change and increasing temperatures.
- The jet stream driving UK weather systems may be getting 'stuck' due to climate change.

Could UK weather patterns be getting stuck?

UK weather systems, driven by winds from the *jet stream*, usually cross from west to east. The jet stream moves north and south but can 'stick' in one position resulting in prolonged periods of the same type of weather, such as heatwaves.

These 'stuck' periods have become more frequent and could be due to climate change.

 Over to you

- Describe **three** points about each of **two** examples of extreme weather in the UK.
- TV, social media and newspapers may report individual weather events as evidence of climate change. Why might this be misleading?

What is the evidence for climate change?

Student Book
See pages 40–1

You need to know:

- the evidence for climate change from the beginning of the Quaternary period to the present day.

What is the evidence for climate change?

We know that climates have changed throughout geological time. For example, scientists using fossil records have found fluctuations in temperature for the last 5.5 million years, and, interestingly, a gradual cooling trend!

Marked fluctuations throughout the last 2.6 million years (the *Quaternary period*) explain *glacial periods* and warmer *inter-glacial periods*. Oxygen trapped in layers of ocean sediments, and water molecules in Antarctic snow, can be analysed to calculate temperature because reliable thermometer records only go back around 100 years.

But these direct measurements indicate a clear warming trend, with most of the increase since the mid-1970s (Figure **1**).

This is 'global warming', which has already had significant effects on global ecosystems and on people's lives.

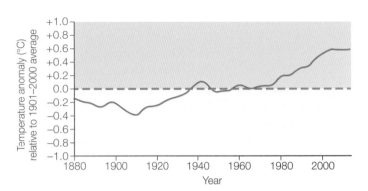

Figure 1 *Average global temperatures (1880–2013) based on recorded temperature records*

Recent evidence for climate change

Shrinking glaciers and melting ice
Some glaciers may disappear by 2035. The extent of Arctic sea ice reached an all-time low in 2014.

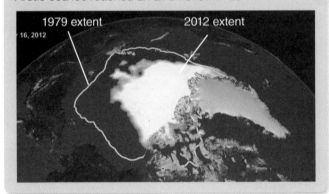

Rising sea level
- Glacier and ice cap melting adds fresh water.
- Thermal expansion – warm ocean waters expand in volume.
- Low-lying islands such as the Maldives, and coastal regions in Bangladesh, India and Vietnam, in danger of flooding.

Seasonal changes
- Tree flowering and bird migration is advancing.
- Bird nesting is earlier than in the 1970s.

 Six Second Summary

- Global temperatures have been cooling gradually over 5.5 million years, but increasing in recent decades.
- Many consider contemporary global warming to indicate climate change.
- Melting glaciers, rising sea levels, changing seasons and direct temperature measures give evidence of climate change.
- Climate change is having a significant effect on global ecosystems and on people's lives.

Over to you

Study Figure **1**.

- Describe the trend of the average temperature throughout the period of the graph.
- Comment on **how far** this is strong evidence for global warming? (Don't forget you'd need to justify your answer with evidence.)

You need to know:

- the natural causes of climate change – orbital changes, solar activity and volcanic activity.

Student Book See pages 42–3

The three main natural causes of climate change

1

Orbital changes – the Milankovitch cycles

Three distinct cycles increase (cooling) or decrease (warming) the distance from the Sun:

- *Eccentricity* – every 100 000 years or so the orbit changes from almost circular, to mildly elliptical (oval) and back again.
- *Axial tilt* – every 41 000 years the tilt of the Earth's axis moves back and forth between 21.5° and 24.5°.
- *Precession (or wobble like a spinning top)* – over a period of around 26 000 years the axis wobbles from one extreme to the other.

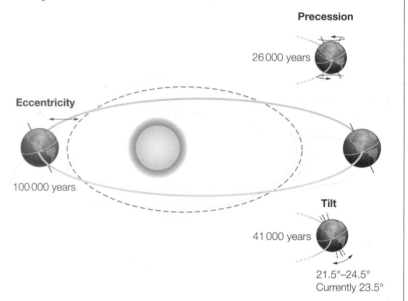

Figure 1 The Milankovitch cycles

2

Solar activity

The surface of the Sun has dark patches called sunspots which mark short-term regions of reduced surface temperature. They are usually accompanied by explosive, high-energy solar flares increasing heat output.

Over a period of around 11 years, sunspots increase from a minimum to a maximum, and back again.

3

Volcanic activity

Volcanic ash can block out the Sun, reducing temperatures on the Earth. This is a short-term impact.

Sulphur dioxide is also blasted out which converts to droplets of sulphuric acid, and acts like mirrors to reflect solar radiation back into space. This longer-term impact (over many years) also reduces temperatures.

 Six Second Summary

- Milankovitch cycles (orbital changes) constantly change the Earth's distance from the Sun.
- Solar activity varies with the number of sunspots and high-energy solar flares.
- Volcanic activity produces ash and sulphuric acid droplets which reduce temperature.

 Over to you

- Make a mnemonic (see page 11) from the **three** natural ways in which climate can change.
- Practise summarising **each** reason why climate can change naturally over time.

You need to know:

Student Book
See pages 44–5

- what the greenhouse effect is
- how human activities can enhance it.

What is the natural greenhouse effect?

The greenhouse effect keeps the Earth naturally warm enough to support life (Figure **1**). It works like a glass greenhouse by:

- greenhouse gases (e.g. water vapour, carbon dioxide (CO_2), methane (CH_4) and nitrous oxides) trapping heat that would otherwise escape into space
- allowing short-wave radiation (light) from the Sun through to the Earth
- trapping some of the longer wavelength radiation (heat) that would otherwise be radiated back into the atmosphere.

Figure 1 *How the greenhouse effect works*

Human impact and the enhanced greenhouse effect

In recent years the amounts of greenhouse gases in the atmosphere have increased (Figure **2**). Scientists believe that this *enhanced greenhouse effect* is due to human activities.

- CO_2 is most important, contributing approximately 60% to the net warming by greenhouse gases.
- Most CO_2 comes from burning fossil fuels in industry and power stations. Transport and farming also contribute.
- Deforestation of tropical rainforests by burning is another major source.
- CH_4 emissions from ever-increasing numbers of farm livestock, rice farming, sewage treatment, and emissions from landfill sites, coal mines and natural gas pipelines are growing even faster than CO_2.

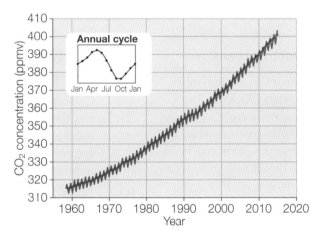

Figure 2 *Increase in CO_2 obtained from direct readings at the Mauna Loa Observatory, Hawaii*

Six Second Summary

- The natural greenhouse effect keeps the Earth warm enough to support life.
- In recent years greenhouse gases produced by human activities have increased.
- This enhanced greenhouse effect is changing climates, weather patterns and sea levels.

Over to you

- Make a mnemonic of the **four** human activities thought to cause an enhanced greenhouse effect.
- The trend of Figure **2** is identical to that of average global temperatures. So does this support the suggestion that human activities may contribute to global warming? You must be able to explain your answer.

You need to know:

- different ways in which the causes of climate change can be managed (mitigated).

Student Book
See pages 46–7

How can climate change be managed?

Alternative energy sources

The burning of **fossil fuels** accounts for 87% of all CO_2 emissions. Alternative sources of energy such as **hydroelectric power (HEP), nuclear power, solar, wind** and **tides** represent **sustainable**, low carbon alternatives.

The UK aims to produce 15% of its energy from **renewable energy sources** by 2020.

Carbon capture

Although not yet economically viable, *carbon capture and storage (CCS)* uses technology to capture CO_2 that is produced by burning fossil fuels in electricity generation and industrial processes. Once captured, the CO_2 is compressed, piped and injected underground for long-term storage in suitable geological reservoirs, such as depleted oil and gas wells.

Figure 1 *Artist's impression of the Hinkley Point nuclear reactor being built in Somerset*

Planting trees

Trees act as carbon sinks, removing CO_2 from the atmosphere by the process of *photosynthesis*. They also release moisture, producing more cloud and so reducing incoming solar radiation.

Tree planting is well established in many parts of the world. In fact, plantations are more efficient at absorbing CO_2 than natural forests.

International agreements

Climate change is a global issue requiring global solutions. Governments are negotiating towards a more sustainable future. For example, the Paris Agreement (2015) was the first legally binding global climate deal. It aims to limit global temperature increases to 1.5 °C above pre-industrial levels.

Global impacts of climate change

- Reduced crop yields and water supplies
- More heat-related illness and disease
- Low-lying coastal areas threatened by flooding
- Changing ecosystems and animal habitats
- More extreme weather events, such as droughts and floods
- Stronger tropical storms
- Desertification

 Six Second Summary

- Alternative energy sources represent sustainable alternatives to fossil fuels.
- Tree planting is established; CCS is not yet economically viable.
- International agreements seek global solutions to issues of climate change.

 Over to you

- Learn two points about **each** way in which climate change could be managed.
- How might you argue the importance of international agreements in helping to solve the problems associated with climate change? Think of **three** things to say.

Student Book
See pages 48–9

You need to know:

- how climate change can be managed by adapting to changes.

How can we adapt to climate change?

Agricultural adaptation

Scientists believe that climate change will have a huge impact on agricultural systems across the world, particularly in low latitudes. To adapt farmers will need to:

- cope with extreme weather such as floods, heatwaves and drought
- manage water supply by storing water, use efficient irrigation systems, grow drought-resistant crops, and adapt to seasonal changes
- plant trees to shade seedlings
- change crops and livestock to suit the new climatic conditions.

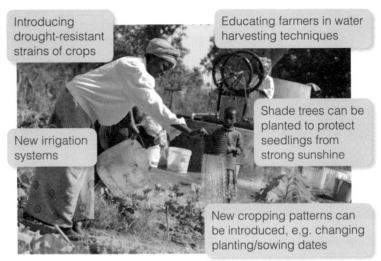

Introducing drought-resistant strains of crops

Educating farmers in water harvesting techniques

New irrigation systems

Shade trees can be planted to protect seedlings from strong sunshine

New cropping patterns can be introduced, e.g. changing planting/sowing dates

Figure 1 *Adapting to climate change – irrigating crops in the Gambia*

Reducing risk from rising sea levels

Having already risen 20 cm since 1900, average sea level rises of up to 1 m by 2100 are possible. This will:

- threaten important agricultural land in countries such as Bangladesh, India and Vietnam
- increase rates of coastal erosion and damage from storm surges
- contaminate freshwater supplies with saltwater.

The low-lying Indian Ocean islands of the Maldives are already tackling this change by adopting practical management strategies (Figure **2**).

Construction of sea walls – a 3 m sea wall is being constructed around the capital Male with sandbags used elsewhere (as in this photo)

Building houses that are raised off the ground on stilts

Restoration of coastal mangrove forests – their tangled roots trap sediment and offer protection from storm waves

Ultimately the entire population could be relocated to Sri Lanka or India

Construction of artificial islands up to 3 m high so that people most at risk could be relocated

Figure 2 *How can the Maldives manage sea-level rise?*

Six Second Summary

- Climate change will have a huge impact on agricultural systems, particularly in low latitudes.
- Farmers will have to adapt by changing crops, livestock and techniques, and manage water supplies.
- Sea-level rise will require management of coastal areas.

Over to you

- Learn **three** threats to farmers posed by climate change.
- Which of the sea-level rise management strategies adopted in the Maldives are immediately relevant, and which are longer-term?

Section B
The living world

Your exam

Section B The living world makes up part of Paper 1: Living with the physical environment.

Paper 1 is a one-and-a-half hour written exam and makes up 35 per cent of your GCSE. The whole paper carries 88 marks (including 3 marks for SPaG) – questions on Section B will carry 25 marks.

You have to study ecosystems and tropical rainforests in Section B. You will then study *either* hot deserts *or* cold environments – in your final exam you will have to answer questions on the three topics you have studied.

> Tick these boxes to build a record of your revision

Your revision checklist

Spec key idea	Theme	1	2	3
5 Ecosystems				
Ecosystems exist at a range of scales and involve the interaction between living and non-living components	5.1 Introducing a small-scale ecosystem			
	5.2 How does change affect ecosystems?			
	5.3 Introducing global ecosystems			
6 Tropical rainforests				
Tropical rainforests have distinctive environmental characteristics	6.1 Environmental characteristics of rainforests			
Deforestation has economic and environmental impacts	6.2 Causes of deforestation in Malaysia			
	6.3 Impacts of deforestation in Malaysia			
Tropical rainforests need to be managed to be sustainable	6.4 Managing tropical rainforests			
	6.5 Sustainable management of tropical rainforests			
7 Hot deserts				
Hot desert ecosystems have distinctive environmental characteristics	7.1 Environmental characteristics of hot deserts			
Development of hot desert environments creates opportunities and challenges	7.2 Opportunities for development in hot deserts			
	7.3 Challenges of development in hot deserts			
Areas on the fringe of hot deserts are at risk of desertification	7.4 Causes of desertification in hot deserts			
	7.5 Reducing desertification in hot deserts			
8 Cold environments				
Cold environments (polar and tundra) have distinctive characteristics	8.1 Characteristics of cold environments			
Development of cold environments creates opportunities and challenges	8.2 Opportunities for development in Svalbard			
	8.3 Challenges of development in Svalbard			
Cold environments are at risk from economic development	8.4 Cold environments under threat			
	8.5 Managing cold environments			

You need to know:

- what is meant by an ecosystem
- components of a small-scale ecosystem
- details of a small-scale ecosystem in the UK.

Example

What is an ecosystem?

An **ecosystem** is a complex natural system made up of plants, animals and the environment. They occur at different scales from small (e.g. pond, woodland) to global (e.g. tropical rainforest). Global-scale ecosystems are called *biomes* – ecosystems at a bigger scale.

Within an ecosystem there are often complex interrelationships (links) between the living (*biotic*) features (e.g. plants, animals and fish) and the (*abiotic*) non-living environmental factors (e.g. climate, soil and light).

- **Producers** (e.g. plants) convert energy from the Sun by *photosynthesis* into carbohydrates (e.g. sugars) for growth.
- **Consumers** get their energy from eating producers, creating direct links within ecosystems (**food chains**) and more complex **food webs**.
- Finally, dead plant and animal material is broken down by **decomposers** (e.g. bacteria and fungi) to add to nutrients within the soil.
- These nutrients are then used by plants in a process called **nutrient cycling**.

A freshwater pond ecosystem

Freshwater ponds provide a variety of habitats (homes) for plants, insects and animals (Figure **1**). You need to learn an example of a producer and a consumer, and also an example of a food chain and of a food web.

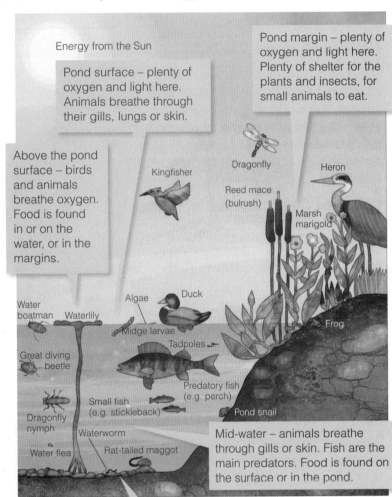

Energy from the Sun

Pond surface – plenty of oxygen and light here. Animals breathe through their gills, lungs or skin.

Pond margin – plenty of oxygen and light here. Plenty of shelter for the plants and insects, for small animals to eat.

Above the pond surface – birds and animals breathe oxygen. Food is found in or on the water, or in the margins.

Kingfisher

Dragonfly

Heron

Reed mace (bulrush)

Marsh marigold

Water boatman   Waterlily

Algae

Duck

Frog

Midge larvae

Tadpoles

Great diving beetle

Predatory fish (e.g. perch)

Pond snail

Dragonfly nymph

Small fish (e.g. stickleback)

Waterworm

Water flea

Rat-tailed maggot

Mid-water – animals breathe through gills or skin. Fish are the main predators. Food is found on the surface or in the pond.

Pond bottom – little oxygen or light. Plenty of shelter (rotting plants and stones) and food. Decomposers and scavengers live here.

Figure 1 *A freshwater pond ecosystem*

Six Second Summary

- An ecosystem is a complex system made up of plants, animals and the environment.
- Ecosystems occur at different scales from small (e.g. pond) to global (biomes).
- There are complex interrelationships between *biotic* and *abiotic* components of ecosystems.
- Within each ecosystem are producers, consumers, a food chain, a food web, decomposers and a nutrient cycle.

Over to you

List the terms that appear **bold** on this page.

Define each term, and illustrate it using examples from a freshwater pond ecosystem.

How does change affect ecosystems?

You need to know:

- how natural changes and human activities affect ecosystems
- an example of such a change.

Student Book See pages 54–5

What are the impacts of change on an ecosystem?

Ecosystems can take thousands of years to develop their sustainable balance. Yet global-scale changes (e.g. climate change) and local-scale changes (e.g. hedgerow removal) can upset this balance.

Natural changes

Slow natural changes have few harmful effects, but rapid changes have serious impacts. For example, extreme weather events like droughts can be devastating to freshwater ponds – killing fish, plants and birds that are dependent upon them.

Changes due to human activities

Human activities can have many impacts on ecosystems. Any one change can have serious knock-on effects (Figure 1).

Agricultural fertilisers can lead to eutrophication: nitrates increase growth of algae, which will deplete oxygen and fish may die.

Woods cut down, destroying habitats for birds and affecting the nutrient cycle.

Ponds may be drained to use for farming. Aquatic plants will die, as will fish and other pond life.

Hedgerows removed to increase size of fields. Habitats will be destroyed, altering the plant/animal balance.

Figure 1 *The impact of human changes on small-scale ecosystems*

Avington Park Lake, Winchester, Hampshire

Over several years, the condition of this historical, ecologically important lake had deteriorated due to lack of maintenance. Silt had accumulated and excessive vegetation blocked a previously impressive view. But birds did thrive.

In 2014 the lake was de-silted and reshaped. New waterside habitats were created to attract nesting birds and waterfowl. The lake is now an attractive, healthy ecosystem supporting a diverse range of wildlife.

Figure 2 *Avington Park Lake restoration*

 Six Second Summary

- Changes to one component can seriously affect the balance of an ecosystem.
- Changes to ecosystems can take place at all scales from local to global.
- Changes to ecosystems can be due to natural causes or human activities.

 Over to you

- In **one** sentence state a) a short-term change and b) a long-term change to ecosystems.
- Practise writing a brief account of Avington Park lake restoration to illustrate positive changes.

Introducing global ecosystems

Student Book
See pages 56–7

You need to know:

- the distribution and characteristics of large-scale global ecosystems (biomes).

The distribution and characteristics of large-scale global ecosystems (biomes)

Biomes are mainly defined by one dominant type of vegetation. They form broad belts usually parallel to lines of latitude (Figure **1**). This is because the climate and characteristics of ecosystems are determined by global atmospheric circulation (see 3.2).

Variations occur in these west-to-east belts of vegetation because of factors such as ocean currents, winds and the distribution of land and sea. These create small changes in temperature and moisture which, in turn, affect the ecosystems.

Tundra – mainly located between the Arctic Circle to about 60°–70° North.

Cold, windy and dry conditions support low-growing plants easily damaged by developments, e.g. oil exploitation and tourism.

Deciduous and coniferous forests – located roughly 50°–60° North. Deciduous trees shed their leaves in winter, but cone-bearing coniferous evergreens are better suited to colder climates and so dominate further north.

Temperate grassland – located 30°–40° north and south of the Equator, and always inland.

Warm, dry summers and cold winters support grasses for grazing animals.

Mediterranean – (also isolated locations south of the Equator). Hot, sunny, dry summers and mild winters support olive groves and citrus fruits.

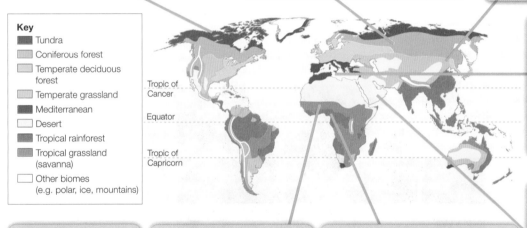

Key
- Tundra
- Coniferous forest
- Temperate deciduous forest
- Temperate grassland
- Mediterranean
- Desert
- Tropical rainforest
- Tropical grassland (savanna)
- Other biomes (e.g. polar, ice, mountains)

Tropic of Cancer

Equator

Tropic of Capricorn

Polar – located around the North and South Poles. Low temperatures (below –50°C) and dry conditions prohibit most plant and animal life.

Tropical grassland (savanna) – located between 15°–30° north and south of the Equator.

Distinct wet and dry seasons support large herds of grazing animals and their predators.

Tropical rainforest – covering 6% of the Earth's land surface mainly close to the Equator

High temperatures and heavy rainfall create ideal conditions for vegetation.

More than half of all plant and animal species, and a quarter of all medicines originate here.

Desert – covering 5% of the Earth's land surface. High daytime temperatures, low night-time temperatures and very low rainfall restrict plants and animals to highly specialised species.

Figure 1 *Global ecosystems (biomes)*

 Six Second Summary

- Large-scale global ecosystems are known as biomes.
- Biomes are defined mainly by the dominant type of vegetation growing there.
- Biomes are distributed in broad belts across the world from west to east, parallel to lines of latitude.
- The climate and characteristics of biomes are determined by global atmospheric circulation.

 Over to you

Make sure that you can explain:
a) why most biomes form broad latitudinal belts across the world
b) why minor variations can occur within them.

You need to know:

- where tropical rainforests are found
- the characteristics of tropical rainforests.

*Student Book
See pages 58–9*

Where are tropical rainforests found?

Tropical rainforests are located mostly a few degrees either side of the Equator between the Tropics of Cancer and Capricorn in the equatorial climate (see 5.3, Figure **1**).

 Big Idea

Tropical rainforests cover 6% of the Earth's surface, yet they support more than 50% of all living organisms!

What are the distinctive characteristics of rainforests?

- Rainforests are hot and humid all year, with high, but variable rainfall. There are no distinct seasons.
- The **biodiversity** is remarkable, supporting more plants and animals than any other biome.
- Tropical rainforests are perfectly adapted to their environment, with distinctive vertical stratification of deciduous trees and climbing plants, which form a dense canopy all competing for light (Figure **1**).

- Soils (*latosols*) are iron-rich and surprisingly infertile because *nutrient cycling* is so rapid. Once tropical rainforests are cleared, their soils are prone to rapid *leaching* where minerals are lost in solution.
- Small changes, such as deforestation or water **pollution**, can have serious knock-on effects on the entire ecosystem.

Lower tree canopy (10–20 m) Shaded, less substantial trees waiting to take advantage of the next available light space. Interlocking spindly branches and climbing woody creepers (*lianas*) form green corridors along which lightweight animals can travel.

Top canopy (35–50 m) Hardy exposed *emergent* trees with straight branchless trunks receive the most light.

Middle canopy (20–35 m) The most productive layer as each mushroom-shaped crown has an enormous photosynthetic surface of dark, leathery leaves. *Drip tips* help them shed water quickly and efficiently.

Figure 1 *Stratification, vegetation and soil adaptations in a tropical rainforest*

Shrub and ground layer (0–10 m) Limited to ferns, woody plants and younger trees because of lack of light. Bacteria and fungi rapidly rot the fallen leaves, dead plants and animals. Thick *buttress* roots help to spread the weight of the towering trees above.

Soils (*latosols*) cycle nutrients rapidly to support new growth. But if the rainforest is cleared they become exposed to excessive leaching and are quickly exhausted of stored nutrients.

 Six Second Summary

- Tropical rainforests are found in a broad band near the Equator.
- The equatorial climate has high rainfall and high temperatures all year.
- Soils (called latosols) in tropical rainforests are surprisingly infertile – they are leached by heavy rainfall.
- Tropical rainforest biodiversity is remarkable, and adapted to a fragile ecosystem which is vulnerable to changes.

Over to you

Draw a spider diagram with the word 'rainforests' at the centre, and add legs to show 'biodiversity', 'climate', 'soils' and 'changes'. Build detail out from each of these.

Student Book
See pages 60–1

You need to know:

- the causes of deforestation in Malaysia.

Example

Deforestation in Malaysia

Malaysia, in south-east Asia, is 67% tropical rainforest. But its rate of **deforestation** is increasing faster than in any tropical country in the world.

Figure 1 *The location of Malaysia*

The threats to Malaysia's rainforests

- *Logging* – in the 1980s, Malaysia became the world's largest exporter of highly valued tropical wood. But destructive *clear felling* has now largely been replaced by **selective logging** of mature trees only (Figure **2**).
- *Road building* – roads are constructed to provide access to logging and mining areas, new settlements and energy projects (Figure **2**).
- *Energy development* – HEP projects boost Malaysia's electricity supplies (Figure **3**).
- **Mineral extraction** – tin mining is established and drilling for oil and gas has recently started.
- *Population pressure* – poor people from urban areas have been encouraged to move into the countryside from rapidly growing cities. This *transmigration* has set up settlements and palm oil plantations.
- **Commercial farming** – Malaysia is the largest exporter of palm oil in the world. Ten-year tax incentives encourage more deforestation for more plantations.
- **Subsistence farming** – traditional short-term clearance is small scale and sustainable, but 'slash and burn' fires can grow out of control destroying large areas of forest.

Figure 2 *Selective logging and road construction in Sarawak, east Malaysia*

Figure 3 *Bakun HEP Dam in Sarawak, which opened in 2011*

Six Second Summary

- 67% of Malaysia is tropical rainforest, and the rate of large-scale deforestation is increasing.
- Threats to Malaysia's rainforests come from logging, energy development, mineral extraction, transmigration, and commercial and subsistence farming.

Over to you

Make a mnemonic (a phrase or sentence, with words beginning with the first letter of each process) to help you remember the list of reasons why deforestation in Malaysia is taking place. Add **one** cause for **each** reason you have listed.

You need to know:

- the impacts of deforestation in Malaysia – soil erosion, loss of biodiversity, contribution to climate change and economic gains and losses.

Student Book
See pages 62–3

Impacts of deforestation in Malaysia

Deforestation destroys the ecosystem and the many habitats that exist on the ground and in the trees (Figure **1**). The stripping of vegetation:

- reduces biodiversity with incalculable losses of undiscovered plant species and their medicinal potential
- exposes the ground (previously shaded, and with soil bound together by the roots of trees and plants) to **soil erosion** by wind and rain
- impacts local and global climates by reducing photosynthesis, transpiration and the cooling effect of evaporation. Consequently there is more CO_2 (which is party responsible for global warming) less moisture to condense into clouds and higher temperatures.

Figure 1 *Rainforest deforestation in Malaysia*

Rainforests versus economic development

Most deforestation is driven by profit. But are these just short-term economic gains?

Economic gains	Economic losses
Job creation – directly in construction and operations, and indirectly in supply and support industries.	Water pollution in an increasingly dry climate may limit supplies.
Tax revenue used to supply public services (e.g. education).	Fires pollute and destroy vast areas of valuable forest.
Improved transport **infrastructure** benefits development and tourism.	Rising temperatures could devastate established farming.
Plantation products support processing industries.	Plants that could form the basis of hugely profitable medicines may become extinct.
HEP is cheap and plentiful.	Climate change could have economic costs (see 4.4 and 4.5).
Minerals are valuable.	Rainforest tourism could decrease.

Six Second Summary

- Deforestation leads to soil erosion, loss of biodiversity and climate change.
- There may be short-term economic gains, but long-term economic losses.

Over to you

Construct a spider diagram to summarise the impacts of deforestation.

You need to know:

*Student Book
See pages 64–5*

- the rates of deforestation worldwide
- how the rates are changing in Brazil
- why all tropical rainforests should be protected.

What are the rates of deforestation?

Every two seconds an area of rainforest the size of a football field is being destroyed – an area the size of China has been lost already! Losses involving landowners and big companies are fastest in Brazil and Indonesia (Figure **1**), but rates of loss are falling in the former and rising in the latter. Brazil's losses are now the lowest on record because:

- Brazil's government has cracked down on illegal deforestation
- Brazil leads the world in **conservation** (protecting over half of the Amazon)
- Brazil is committed to reducing carbon emissions to tackle climate change
- consumer pressure not to use products from deforested areas has led to a decline in cattle ranching

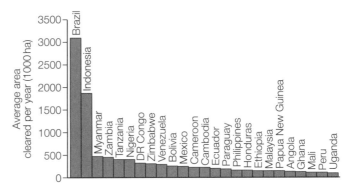

Figure 1 Rates of tropical rainforest deforestation, 2010–15

Why should tropical rainforests be protected?

Biodiversity
Tropical rainforests contain more than half of all the plants and animals in the world – and how many species are still to be discovered?

Medicine
Around 25% of all medicines come from rainforest plants

Climate change
Rainforests absorb and store CO_2

Resources
Valuable hardwoods, nuts, fruit and rubber

Climate
Rainforests prevent the climate from becoming too hot and dry – and produce 28% of the world's oxygen!

Water
Rainforests are important sources for clean water

People
Indigenous tribes live sustainably, e.g. the Achuar tribe of the Peruvian Amazon

 Six Second Summary

- Tropical rainforests globally are being destroyed at different rates.
- Brazil has the fastest rates of deforestation, but this rate has fallen dramatically.
- Important reasons why rainforests should be protected include biodiversity, climate change, resources and threats to indigenous people.

Over to you

Practise outlining a) the global pressures on rainforests, b) why tropical rainforests should be protected.

How can rainforests be managed sustainably?

Rainforests need to be managed **sustainably** in order to:

- ensure that they remain a lasting resource for future generations
- harness valuable resources without causing long-term damage to the environment.

Indigenous tribes like the Peruvian Achuar manage rainforests sustainably, but support relatively few people. It is the wealthy landowners, large companies and illegal loggers, in their drive for profits that do the damage.

Sustainable commercial management

- *Selective logging and replanting* – introduced in Malaysia (Figure **1**) – avoids the completely destructive *clear felling*.
- *Conservation and education* encourages preservation of rainforests in national parks and nature reserves for scientific research and tourism (e.g. in Brazil).
- **Ecotourism**, such as in Costa Rica and Malaysia, introduces people to the natural world and provides long-term income to local people and governments (Figure **2**).
- *International agreements* recognise the global importance of rainforests in combating climate change. They include 'debt-for-nature-swapping' – agreements whereby some donor countries and organisations reduce their debt repayment demands in return for a halt to deforestation.

The Forest Stewardship Council (FSC) promotes sustainably managed forestry through education programmes and its FSC-labelled products.

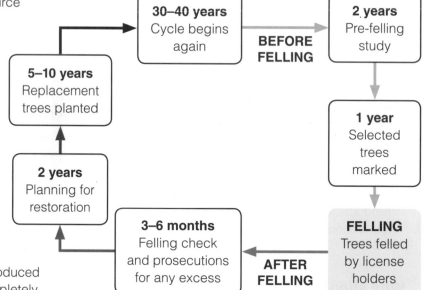

Figure 1 Malaysia's Selective Management System

Figure 2 Tourist accommodation in an eco-lodge

 Six Second Summary

- Rainforests need to be managed sustainably in order to preserve them for the future and to make use of their resources without damaging the environment.
- Rainforests can be managed sustainably by selective logging, replanting, conservation, education, ecotourism and international agreements.

 Over to you

Take four cards and label each one 'Selective logging and replanting', 'Conservation and education', 'Ecotourism' and 'International agreements'. On each card, outline how and why it contributes to sustainable management of rainforests. Then arrange the cards in order of importance.

Environmental characteristics of hot deserts

You need to know:

- the location of hot deserts and their climatic characteristics
- how plants and animals adapt.

Student Book
See pages 68–9

Where are hot deserts found?

Deserts are dry (arid) areas. They are both hot (e.g. the Sahara) and cold (e.g. Antarctica), but all receive less than 250 mm of rainfall per year.

Hot deserts are mostly found in a belt between approximately 30°N and 30°S (Figure **1**). This is largely explained by global atmospheric circulation (see 3.1). Most occupy dry continental interiors, but coastal deserts also exist (e.g. the Atacama Desert in South America).

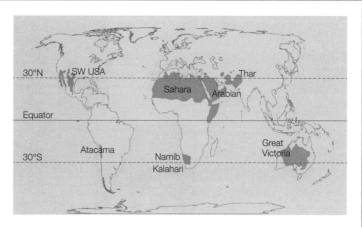

Figure 1 *Location of the world's hot deserts*

What are the characteristics of hot deserts?

- Very low rainfall.
- Extreme *temperature range*. Lack of cloud cover allows high daytime temperatures, but very cold nights.
- Soils tend to be sandy or stony. Limited leafy vegetation means little organic matter and fertility.
- Soils are *saline* (salty) because evaporation of moisture draws salts to the surface.
- A wide *diversity* of plants, animals and birds find ways to survive. Vegetation has several adaptations (Figure **2**). Nocturnal rodents live in burrows underground. Snakes and lizards have waterproof skins which retain moisture, and camels can withstand days without water.

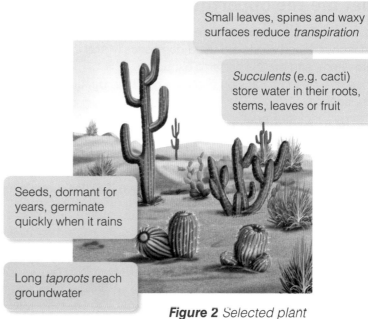

Small leaves, spines and waxy surfaces reduce *transpiration*

Succulents (e.g. cacti) store water in their roots, stems, leaves or fruit

Seeds, dormant for years, germinate quickly when it rains

Long *taproots* reach groundwater

Figure 2 *Selected plant adaptations to the hot desert climate*

Six Second Summary

- Hot deserts are located in a belt between 30°N and 30°S.
- The climate has very low rainfall and an extreme temperature range.
- Soils are dry and infertile with little organic matter.
- A wide diversity of plants, animals and birds have adapted to the hostile environment.

Over to you

Checklist.
Can you:
- locate and name examples of hot deserts?
- explain their aridity and extreme temperatures?
- describe how plants and animals adapt to the hostile environment?

You need to know:

- how people use hot desert environments
- what opportunities there are for economic development in the Thar Desert.

Student Book
See pages 70–1

Where is the Thar Desert?

The Thar Desert is the most densely populated desert in the world! It stretches across north-west India and into Pakistan.

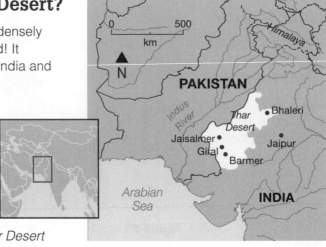

Figure 1 *Location of the Thar Desert*

Big Idea

Hot deserts are so much more than empty, useless wastelands!

What are the opportunities for development?

- *Mineral extraction* – valuable reserves of gypsum, feldspar, phospherite and kaolin are used domestically and for export. Limestone and marble are also extracted.
- *Tourism* – desert safaris on camels exploit the beautiful landscapes.
- *Energy* – includes coal, oil, solar and wind (Figure **2**).
- *Farming* – subsistence grazing of animals in grassy areas, vegetable and fruit cultivation, and commercial farming of pulses, sesame, mustard, maize, cotton and wheat (Figure **3**).
- **Irrigation** from the Indira Gandhi Canal (see 7.3) has made what was once scrubby desert productive.

Figure 2 *The Jaisalmer Wind Park – India's largest wind farm*

Figure 3 *Growing wheat on irrigated land in the desert*

 **Six Second Summary**

- The Thar Desert offers opportunities for economic development, including mineral extraction, tourism, and wind and solar energy.
- Improvements in irrigation have led to the growth of commercial farming.

 Over to you

Summarise this page in **five** bullet points that disprove the assumption that hot deserts are empty, useless wastelands.

Student Book
See pages 72–3

You need to know:

• the challenges of economic development in the Thar Desert.

Example

Extreme temperatures

The Thar Desert suffers from extremely high temperatures which:

• makes physical work hard, especially for those who work outside, such as farmers
• causes high rates of evaporation leading to water shortages
• determines plant and animal adaptations (see 7.1) and requires shade for livestock.

Water supply

Water in the Thar Desert is a scarce resource because:

• annual rainfall is low, and high temperatures and strong winds cause high evaporation
• as population has increased, farming and industry have developed, increasing the demand for water.

There are several sources:

• traditional storage ponds – human-made *johads* and natural *tobas* (Figure **1**)
• a few intermittent rivers and streams
• underground aquifers requiring wells, though often the water is saline (salty).

This is why the Indira Gandhi Canal is so important (Figure **2**). Constructed in 1958 with a length of 650 km, it has helped to transform an extensive area of desert and revolutioned farming.

Figure 1 Collecting water in the desert

Commercial farming of crops such as wheat and cotton, now flourishes in an area that used to be scrub desert.

The canal provides drinking water to many people in the desert.

Accessibility

Vast barren areas and very extreme weather limit the road network, as tarmac can melt during the day and strong winds blow sand over the roads. Many places are accessible only by camel.

Two of the main areas to benefit from the canal are centred on the cities of Jodhpur and Jaisalmer where over 3500 km² of land is under irrigation.

Figure 2 The Indira Gandhi Canal

 Six Second Summary

• Extreme temperatures, water shortages and poor accessibility present challenges for development in the Thar Desert.
• The Indira Gandhi Canal is the main source of irrigation, and has revolutionised farming.

Over to you

Why might many people argue that very high temperatures and a limited road network present the greatest challenges for development in the Thar Desert?

Student Book
See pages 74–5

You need to know:

- the causes of desertification in hot deserts.

What is desertification?

Desertification is where land is gradually turned into desert. It occurs mostly on the ecologically fragile borders of existing deserts (Figure 1). Desertification:

- is a result of both natural (e.g. droughts) and human (e.g. mismanagement) events
- affects both poor and rich countries
- threatens one billion people in areas at risk.

Slight changes in temperature and rainfall (associated with climate change) make these areas even more prone to **over-cultivation, overgrazing** or the stripping of vegetation for fuelwood.

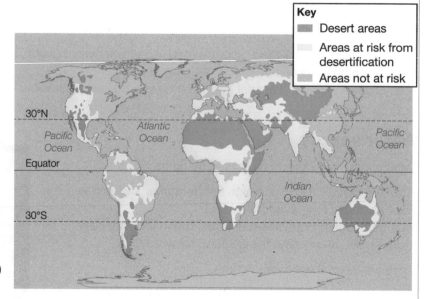

Key
- Desert areas
- Areas at risk from desertification
- Areas not at risk

30°N — Pacific Ocean — Atlantic Ocean — Pacific Ocean

Equator

Indian Ocean

30°S

Figure 1 *Areas at risk of desertification*

Causes and consequences of desertification

Climate change is resulting in drier conditions and unreliable rainfall in some regions (e.g. the Sahel on the southern margins of the Sahara).

Soil erosion Where vegetation is destroyed, exposing soil which cracks and breaks up, making it vulnerable to erosion by wind and rain.

Salinisation Rapid evaporation of poorly practised irrigation leads to surface salts building-up, which kill the plants (see 7.5, Figure 1).

Overgrazing Population pressure results in the limited vegetation supporting too many animals (e.g. nomadic Bedouin herding more sheep, goats and camels in the Badia, eastern Jordan).

Over-cultivation More people need more food, which exhausts the soil turning it to dust.

Fuelwood Population growth increases demand. Trees that are stripped of branches eventually die.

 Six Second Summary

- Desertification is the process of deserts spreading.
- It is the result of both natural events and human mismanagement.
- The ecologically fragile borders of existing deserts are most at risk.

Over to you

- List the causes of desertification using headings 'Natural causes' and 'Human mismanagement'. Add examples where you can.
- Consider which of these is more important in explaining desertification.
- Could you justify your answer?

Student Book
See pages 76–7

You need to know:

- how desertification can be reduced.

Holding back the desert

Land at risk from desertification needs to be managed sustainably.

- *Water and soil management* – irrigation needs to be managed carefully if salinisation is to be avoided. Too much irrigation and/or badly drained schemes build-up toxic salts on the surface (Figure **1**).
- *Ponding banks* – low walls enclose areas of land to store water.
- *Contour traps* – embankments built along slopes prevent soil being washed away during heavy rainfall (e.g. Australia and the Badia, Jordan).
- *National park status* – gives legal protection to areas at risk (e.g. the Desert National Park in the Thar Desert, India).
- *Tree planting* – reduces soil erosion because the roots bind the soil together, and the leaves and branches provide shade, grazing for animals and fuelwood (e.g. *prosopis cineraria* in the Thar Desert).

Appropriate technology

Practical and sustainable approaches to farming address the needs of poor people who are unable to afford expensive machinery.

For example, 'magic stones' in Burkina Faso, West Africa have reduced desertification and increased crop yields by up to 50 per cent by trapping water and soil (Figure **2**).

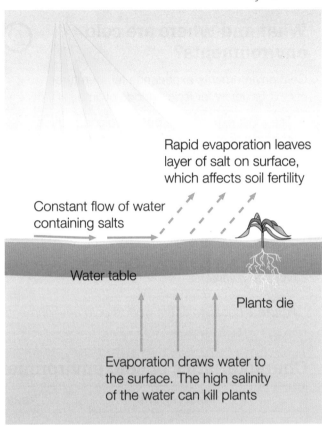

Rapid evaporation leaves layer of salt on surface, which affects soil fertility

Constant flow of water containing salts

Water table

Plants die

Evaporation draws water to the surface. The high salinity of the water can kill plants

Figure 1 *The process of salinisation*

Figure 2 *Walls of 'magic stones' in Burkina Faso*

Six Second Summary

- Areas at risk from desertification need to be managed sustainably.
- Sustainable management includes effective water and soil management, national park protection, tree planting and the use of appropriate technology.

Over to you

- List **four** ways of sustainable management to help stop desertification.
- Make sure that you understand the meaning and importance of appropriate technology, and can quote examples.

Student Book
See pages 78–9

You need to know:

- the location of cold environments
- climatic, soil, plant and animal characteristics of cold environments
- vegetation adaptations associated with hostile conditions.

What and where are cold environments?

Cold environments experience temperatures of 0°C or below for long periods of time.

- The most extreme are **polar** regions and ice sheets where temperatures are below zero all year (Figure **1**).
- *Tundra* areas (around the northern Arctic circle) are less extreme though soils are permanently frozen (**permafrost**) and summers short.
- Less extreme are Alpine areas of high mountains, which simply experience very cold winters.

Figure 1 *Location of the world's cold environments*

 **Big Idea**

Cold environments cover one-third of the world's land surface. They are ecologically fragile and present demanding development and conservation challenges.

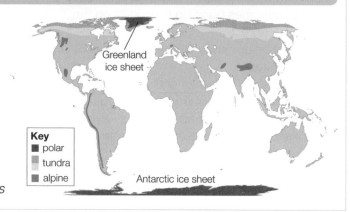

Greenland ice sheet

Key
- ■ polar
- ■ tundra
- ■ alpine

Antarctic ice sheet

Characteristics of cold environments

	Polar regions	**Tundra**
Climate	Extreme – winter temperatures can fall below –50°C, strong winds, low snow totals	Less extreme – winter temperatures may drop to –20°C, warm brief summers, higher precipitation (snow) in coastal regions
Soils	Permafrost covered by ice	Surface permafrost melting in summer causing waterlogging
Plants	Few mosses and lichens on the fringes of the ice	Low-growing, flowering plants with special adaptations (Figure **2**)
Animals	Polar bears well adapted with thick fur and foot pads; Antarctic penguins rear their young on the land	Several species due to more food options and less extreme climate (e.g. Arctic fox and Arctic hare); summer birds (e.g. ptarmigans) and insects (e.g. midges and mosquitoes)

Low-growing so protected from strong winds

Thick bark stems improve stability

Small leathery leaves retain moisture

Hairy stems insulate the plant

Bright red berries attract birds to help spread the seeds

Figure 2 *How the bearberry plant adapts to the tundra environment*

 **Six Second Summary**

- Cold environments have very low temperatures and strong winds.
- Polar regions are most extreme and arid (low snow totals), with tundra environments less so.
- Plants and animals have adapted to the hostile conditions.

 **Over to you**

Make sure that you can explain the distinctive characteristics of plants and soils in the less extreme tundra environment.

- the location of Svalbard
- development opportunities in mineral extraction, energy, fishing and tourism.

*Student Book
See pages 80–1*

Example

Where is Svalbard?

Norway's Svalbard is close to the Mid-Atlantic Ridge and is the world's most northerly inhabited territory (Figure **1**).

Svalbard has five major islands, 60% of which are covered in glaciers and the rest of the land is tundra. There are no trees – it is too cold!

Most of its population of around 2700 live in the main town of Longyearbyen on Spitzbergen, the largest of the islands.

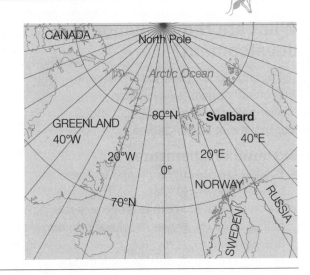

Figure 1 *Location of the Norwegian territory of Svalbard*

What are the opportunities for development in Svalbard?

- *Mineral extraction* – rich reserves of coal, which is the main economic activity but environmentally controversial.
- *Energy developments* – the Longyearbyen coal-fired power station supplies all of Svalbard's energy needs (Figure **2**). Carbon capture and storage (see 4.4) is a likely development in the future. *Geothermal energy* is a likely future source of energy,
- *Fishing* – the Arctic waters of the Barents Sea are rich fishing grounds with 150 species, including cod, herring and haddock. Fishing is carefully controlled and monitored to ensure sustainability of the ecosystem.
- *Tourism* – now provides as many jobs as mining. Increasingly popular in recent years as tourists seek the Northern Lights and explore more extreme natural environments.
  - Cruise passengers land at Longyearbyen seeking glaciers, fjords and wildlife – especially polar bears
  - Adventure tourists seeking hiking, kayaking and snowmobile safaris (Figure **3**).

Figure 2 *The coal-fired power station at Longyearbyen*

Figure 3 *Adventure tourism in Svalbard*

- Svalbard is a group of islands in the Arctic Ocean, close to the Mid-Atlantic Ridge.
- Development opportunities include coal mining, geothermal energy, fishing and tourism.

- Define the phrase 'environmentally controversial'.
- Learn **two** points to explain why coal is controversial; then **two** more explaining why fishing and tourism are also controversial.

Example

*Student Book
See pages 82–3*

You need to know:

- the challenges for development in Svalbard.

What are the challenges for development?

Living and working in such an extreme environment poses extraordinary challenges, not least throughout the four months of winter darkness.

Extreme temperatures

- Even in Longyearbyen, winter temperatures can fall below −30 °C.
- Given the risk of frostbite, several layers of thick clothes, gloves, socks and boots are essential.
- Protected like this, outside work can be slow, difficult and dangerous.

Construction

- Most building, construction and maintenance happen during the brief summer.
- The frozen ground surface (permafrost) has to be protected from melting, or buildings would collapse.
- Most dirt and gravel roads are raised above the ground surface.

Services

- Most power, water and sanitation pipes have to be heated, insulated and raised above ground (Figure **2**).
- This allows easy maintenance and prevents thawing of the permafrost.

Accessibility

- Svalbard can only be reached by sea or air.
- There are no roads outside Longyearbyen.
- International flights link to mainland Norway and Russia, with smaller aircraft connecting to other islands.
- Most people use snowmobiles, particularly in winter (Figure **3**).

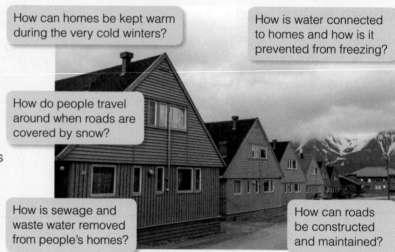

How can homes be kept warm during the very cold winters?

How is water connected to homes and how is it prevented from freezing?

How do people travel around when roads are covered by snow?

How is sewage and waste water removed from people's homes?

How can roads be constructed and maintained?

Figure 1 *Challenges of living and working in Svalbard*

Figure 2 *Overground service pipes*

Figure 3 *Snowmobiles parked in Longyearbyen*

Six Second Summary

- Living and working in the extreme environment of Svalbard poses development challenges.
- These include inaccessibility, four months of darkness, extreme cold, difficulties with construction and service provision on permafrost.

Over to you

- Describe **four** challenges of development in Svalbard.
- Explain **one** way in which **each** challenge is overcome.

*Student Book
See pages 84–5*

You need to know:

- how cold environments are at risk from economic activities
- why these wilderness areas need protecting.

Why are cold environments fragile?

Tundra vegetation takes a very long time to become established. Tundra is a delicate ecosystem which is easily disturbed by human activities. For example, off-road driving in summer leaves deep tyre tracks scaring the swampy, thawed, surface permafrost. Recovery can take decades!

The risks of economic development

Rich reserves of oil and gas remain in high demand, but their exploitation requires the construction of:

- access roads through forests and across tundra vegetation
- supply bases and settlements for workers
- drilling equipment and pipelines.

The potential for long-lasting, if not permanent damage to such **fragile environments** is great, especially from pollution incidents (Figure **1**).

Trees killed by the oil spill

Electricity pylons will have resulted in tree clearance and environmental damage

Risk of fire, either started deliberately or by a lightning strike

Oil has leaked from this broken pipeline

River has become polluted and is now totally lifeless

River edge habitats polluted and destroyed – the vegetation may never recover

Figure 1 *Oil-polluted river in Siberia, Russia*

Why cold environments need protecting

Important unpolluted, unspoilt, outdoor laboratories for scientific research (e.g. climate change)

Many indigenous people (e.g. Arctic Inuit) depend on the wildlife and survive by hunting and fishing

Wild beauty and potential for adventure activities attracts tourism which benefits their economies

Opportunities for forestry and fishing

Home to a rich variety of birds, animals and plants

 Six Second Summary

- Cold environment ecosystems are easily damaged and can take a very long time to recover.
- Exploitation of oil and gas reserves can cause pollution.
- There are powerful reasons for protecting these fragile environments.

Over to you

Write **three** questions about the key themes of this page to test a friend. Make sure you are able to write the answers as well!

You need to know:

- strategies to reduce the risks to cold environments.

Student Book
See pages 86–7

How can risks to cold environments be reduced?

Cold environments need to be managed sustainably in three ways.

1

Using technology – the trans-Alaskan pipeline

Discovering oil in Prudhoe Bay, northern Alaska, in 1969 stimulated a technological solution to Arctic sea ice preventing tanker movements in winter – a revolutionary 1300 km pipeline. The trans-Alaskan pipeline (Figure **1**):

- crosses two mountain ranges and 800 rivers
- is raised and insulated to prevent the hot oil melting the permafrost, while allowing caribou to migrate underneath (Figure **2**)
- is engineered to slide during earthquakes, but with automatic shut-off systems if there is a leak.

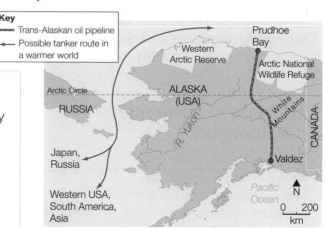

Figure 1 *Route of the trans-Alaskan pipeline and protected wilderness areas*

2

Action by governments

Various laws and policies protect Alaska's unique environment, native people and marine habitats. These include the US government managing protected **wilderness areas** and their abundant wildlife.

Such wilderness protection has even been achieved on an international scale. For example, the Antarctic Treaty goes further by:

- preventing economic development
- promoting scientific research
- controlling tourism to keep disturbance at a minimum.

Figure 2 *The trans-Alaskan pipeline*

3

Conservation groups

Conservation groups such as the World Wide Fund for Nature (WWF, originally World Wildlife Fund) are working with oil companies, Inuit organisations, local communities and government regulators to plan for a sustainable future for the Arctic.

Figure 3 *Antarctica – the world's last great wilderness*

 Six Second Summary

- Cold environments need to be managed sustainably.
- Technology can be used to reduce the impact of development.
- Action by conservation groups, governments and international agreements help protect cold environments, manage development and raise awareness.

 Over to you

Make a list of the arguments **for** and **against** protecting wilderness areas from economic development.

Section C
Physical landscapes in the UK

Your exam

Section C Physical landscapes in the UK makes up part of Paper 1: Living with the physical environment.

Paper 1 is a one-and-a-half hour written exam and makes up 35 per cent of your GCSE. The whole paper is carries 88 marks (including 3 marks for SPaG) – questions on Section C will carry 30 marks.

You have to study the introduction to UK physical landscapes. You must also study two of the other three types of landscapes – in your final exam you will have to answer any two questions from a choice of three.

Your revision checklist

Tick these boxes to build a record of your revision

Spec key idea	Theme	1	2	3
9 UK physical landscapes				
The UK has a range of diverse landscapes	9.1 The UK's relief and landscapes			
10 Coastal landscapes in the UK				
The coast is shaped by a number of physical processes	10.1 Wave types and their characteristics			
	10.2 Weathering and mass movement			
	10.3 Coastal erosion processes			
Distinctive coastal landforms are the result of rock type, structure and physical processes	10.4 Coastal erosion landforms			
	10.5 Coastal deposition landforms			
	10.6 Coastal landforms at Swanage (1)			
	10.7 Coastal landforms at Swanage (2)			
Different management strategies can be used to protect coastlines from the effects of physical processes	10.8 Managing coasts – hard engineering			
	10.9 Managing coasts – soft engineering			
	10.10 Managing coasts – managed retreat			
	10.11 Coastal management at Lyme Regis			
11 River landscapes in the UK				
The shape of river valleys changes as rivers flow downstream	11.1 Changes in rivers and their valleys			
	11.2 Fluvial (river) processes			
Distinctive fluvial (river) landforms result from different physical processes	11.3 River erosion landforms			
	11.4 River erosion and deposition landforms			
	11.5 River landforms on the River Tees			
Different management strategies can be used to protect river landscapes from the effects of flooding	11.6 Factors increasing flood risk			
	11.7 Managing floods – hard engineering			
	11.8 Managing floods – soft engineering			
	11.9 Managing floods at Banbury			
12 Glacial landscapes in the UK				
Ice was a powerful force in shaping the physical landscape of the UK	12.1 Processes in glacial environments			
Distinctive global landforms result from different physical processes	12.2 Glacial erosion landforms			
	12.3 Glacial transportation and deposition landforms			
	12.4 Glacial landforms at Cadair Idris			
Glaciated upland areas provide opportunities for different economic activities, and management strategies can be used to reduce land use conflicts	12.5 Economic opportunities in glaciated areas			
	12.6 Conflict in glaciated areas			
	12.7 Managing tourism in the Lake District			

Student Book
See pages 90–1

You need to know:

- about the location of upland and lowland areas in the UK
- about the UK's river systems.

Relief, rocks, landscape

 Big Idea

Relief describes the physical features of the landscape. It includes:

- height above sea level
- steepness of slopes
- the shape of landforms.

Relief depends greatly on *geology* or rock type. Figure **1** shows the 'Tees-Exe line' joining the River Tees in north-east England with the River Exe in the south-west.

- To the north and west are the uplands of England, Wales and Scotland. More resistant rocks, such as granite and slate, are found here.
- South and east of the line are the lowlands of central and southern England. Weaker rocks such as clays and limestone form low-lying plains and more rolling landscapes.

A **landscape** is an area whose character is the result of the action, and interaction of natural and human factors. The UK has a varied landscape, and much of it is determined by geology.

River systems

The UK has an extensive river system. Most rivers have their source in mountain ranges or hills and flow to the sea. The River Severn, for example, rises in the Cambrian Mountains in Wales, is joined by the River Avon and flows into the Bristol Channel.

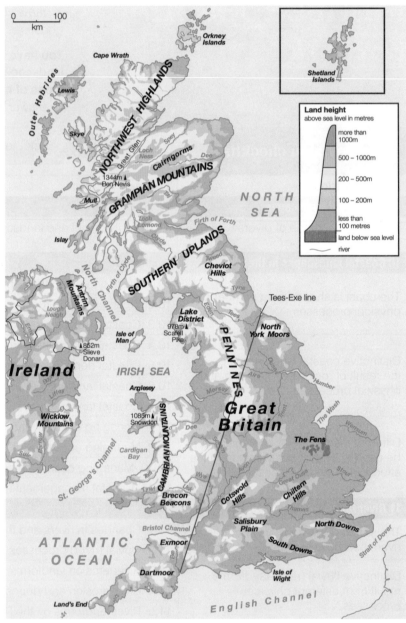

Figure 1 *Atlas map of the UK*

 Six Second Summary

- Relief describes the height, steepness and shape of the landscape.
- Relief is determined by geology.
- The geology and relief of the northern and western UK is very different from that of the southern and eastern UK.
- Landscapes result from the interaction between natural and physical factors.

 Over to you

Draw a spider diagram to show the factors that influence landscapes in the UK. Explain how different types of rock determine the UK's landscapes.

You need to know:

- how waves form
- what happens when they reach the coast
- about different types of waves.

*Student Book
See pages 92–3*

How do waves form?

- By wind blowing over the sea.
- Friction with the surface of the water causes ripples that develop into waves.
- *Tsunamis* form when earthquakes or volcanic eruptions shake the seabed.

The distance that wave-generating winds blow across the water is called the *fetch*.
Remember: The longer the fetch, the bigger the wave.

What happens when waves reach the coast?

In the open sea there is little horizontal movement of water. Figure **1** shows what happens as waves approach the shore.

Figure 1 *Waves approaching the coast*

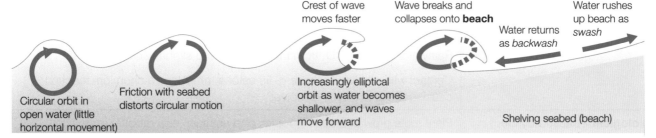

Crest of wave moves faster

Wave breaks and collapses onto **beach**

Water returns as *backwash*

Water rushes up beach as *swash*

Circular orbit in open water (little horizontal movement)

Friction with seabed distorts circular motion

Increasingly elliptical orbit as water becomes shallower, and waves move forward

Shelving seabed (beach)

Two types of wave

1

Constructive waves

Formed by storms often hundreds of kilometres away. Common in summer.

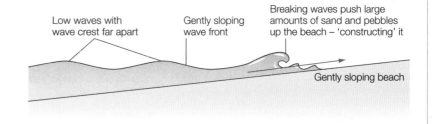

Low waves with wave crest far apart

Gently sloping wave front

Breaking waves push large amounts of sand and pebbles up the beach – 'constructing' it

Gently sloping beach

Figure 2 *Constructive waves*

2

Destructive waves

Formed by local storms close to the coast. Common in winter.

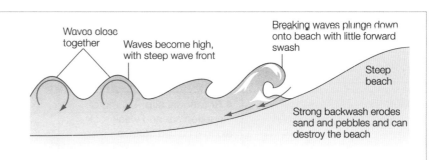

Waves close together

Waves become high, with steep wave front

Breaking waves plunge down onto beach with little forward swash

Steep beach

Strong backwash erodes sand and pebbles and can destroy the beach

Figure 3 *Destructive waves*

 **Six Second Summary**

- Most waves are caused by the friction of the wind on the sea.
- There are two types of wave – constructive and destructive.

 **Over to you**

Draw **two** spider diagrams – one for constructive and one for destructive waves. Add legs for their different characteristics: wave height; wave length; swash; backwash; building/destroying the beach; and anything else you can think of.

You need to know:

- how the processes of weathering and mass movement combine with the action of the waves in shaping the coast.

Student Book
See pages 94–5

Why do cliffs collapse?

Because of weathering – the weakening or decay of rock due to the action of weather, plants and animals.

Type of weathering	Example and description
1 Mechanical (physical) – the disintegration of rock	*Freeze-thaw* Water collects in cracks in rock. At night, water freezes and expands, making cracks larger. As temperature rises, ice thaws and water seeps deeper into rock. Repeated freezing and thawing makes rock fragments break off. They collect as scree at the cliff foot.
2 Chemical – caused by chemical changes	*Carbonation* Rainwater absorbs CO_2 from the air becoming slightly acidic. Contact with alkaline rocks, e.g. limestone, produces a chemical reaction causing rocks to slowly dissolve.
3 Biological – caused by the actions of flora and fauna	Plant roots grow in cracks in rocks, and animals (e.g. rabbits) burrow into weak rocks

What is mass movement?

It's the downward movement (**sliding**) of weathered material and rock under the influence of gravity. Figure **1** shows some of the types of mass movement found at the coast.

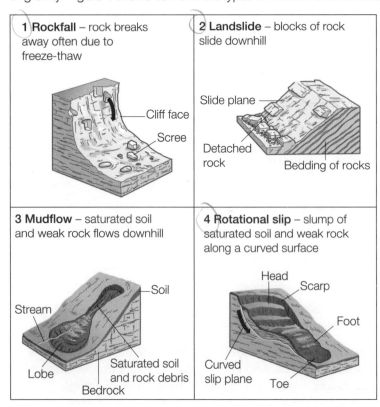

1 **Rockfall** – rock breaks away often due to freeze-thaw
Cliff face
Scree

2 **Landslide** – blocks of rock slide downhill
Slide plane
Detached rock
Bedding of rocks

3 **Mudflow** – saturated soil and weak rock flows downhill
Soil
Stream
Lobe
Saturated soil and rock debris
Bedrock

4 **Rotational slip** – slump of saturated soil and weak rock along a curved surface
Head
Scarp
Foot
Curved slip plane
Toe

Figure 1 *Types of mass movement at the coast*

⏱ **Six Second Summary**

- Weathering is the weakening and breakdown of rock. **Three** main types are mechanical (physical), chemical and biological.
- Mass movement is the downward movement of material. **Four** main examples are rockfall, landslide, mudflow and rotational slip.

 **Over to you**

From memory, draw a diagram to show the process of freeze-thaw weathering. Add detailed labels.

Student Book
See pages 96–7

You need to know:

- how the processes of erosion and deposition combine with the action of the waves to shape the coast.

What is coastal erosion?

Erosion means wearing away the landscape. The processes of coastal erosion are shown in the table.

1 Solution	Dissolving of soluble chemicals in rock, e.g. limestone
2 Corrasion	Rock fragments picked up by the sea are thrown at the cliff. They scrape and wear away the rock.
3 Abrasion	The 'sandpapering' effect of pebbles grinding over a rocky platform.
4 Attrition	Rock fragments carried by the sea knock against each other becoming smaller/more rounded.
5 Hydraulic power	The power of the waves as they hit a cliff. Trapped air is forced into cracks in the rock eventually causing it to break up.

Figure 1 *Processes of coastal erosion*

Why is sediment deposited?

Deposition happens when water slows down and waves lose their energy.

- Beaches are formed of sediment deposited in bays.
- Mudflats and saltmarshes are often found in sheltered estuaries behind spits.

Six Second Summary

- There are **five** processes of coastal erosion.
- There are **four** ways sediment is transported along the coast.
- Longshore drift moves sediment along the coast.
- Deposition happens when waves lose their energy.

Over to you

From memory, draw a diagram to show the **five** processes of coastal erosion. Add annotations to explain the processes.

How is sediment transported?

Sediment transport occurs in four different ways – see the diagram.

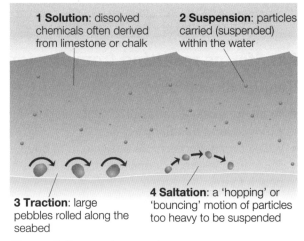

Figure 2 *Types of coastal transport*

Longshore drift

The movement of sediment depends on the direction that waves approach the coast, as a result of the prevailing wind direction.

1 Where waves approach 'head on' sediment moves up and down the beach.

2 Where waves approach at an angle, sediment moves along the beach in a zigzag pattern. This is called **longshore drift**.

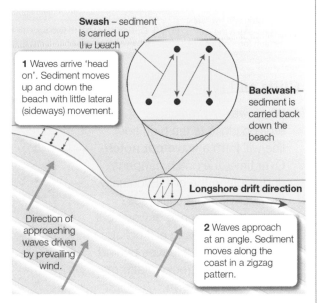

Figure 3 *Longshore drift*

*Student Book
See pages 98–9*

- that a landform is a natural feature formed by the processes of erosion, transportation and deposition
- about the characteristics and formation of coastal landforms.

What factors influence coastal landforms?

1 *Rock type* – some rocks (e.g. granite, limestone) are tougher and more resistant to erosion than others. Softer rocks (e.g. clays, sands) are more easily eroded.

2 *Geological structure* – includes the way that rock has been *folded* or tilted. *Faults* (cracks) form lines of weakness.

Landforms resulting from erosion

Headlands and bays

- Tougher, resistant bands of rock are eroded slowly to form **headlands**.
- Weaker rock erodes more easily to form **bays**. Bays are sheltered, deposition occurs, and a beach forms.

Figure 1 *Formation of headlands and bays*

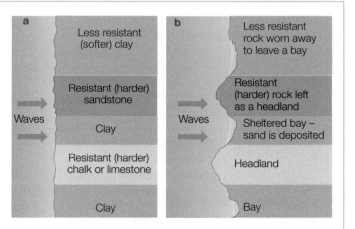

Caves, arches and stacks

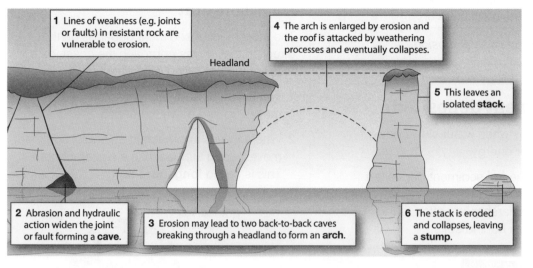

1 Lines of weakness (e.g. joints or faults) in resistant rock are vulnerable to erosion.

4 The arch is enlarged by erosion and the roof is attacked by weathering processes and eventually collapses.

Headland

5 This leaves an isolated **stack**.

2 Abrasion and hydraulic action widen the joint or fault forming a **cave**.

3 Erosion may lead to two back-to-back caves breaking through a headland to form an **arch**.

6 The stack is eroded and collapses, leaving a **stump**.

Figure 2 *Formation of caves, stacks and arches*

Cliffs and wave-cut platforms

- When waves break against a **cliff**, erosion close to the high tide line will form a **wave-cut notch**. Over time, the notch deepens, undercutting the cliff. Eventually the overlying cliff collapses.
- Through a sequence of wave-cut notch formation and cliff collapse, the cliff retreats. It leaves behind a gently sloping rocky platform – a **wave-cut platform**.

 Six Second Summary

- Coastal landforms are influenced by rock type and geological structure.
- Different types of rock erode at different rates.
- Coastal erosion produces distinctive landforms.

Over to you

- Close your book and name **five** coastal erosion landforms.
- Sketch an annotated diagram to show how **two** of them form.

*Student Book
See pages 100–1*

You need to know:

- about the characteristics and formation of landforms resulting from coastal deposition.

Beaches

Beaches are deposits of sand and shingle.

- Sandy beaches are mainly found in sheltered bays and are created by constructive waves.
- Along high-energy coasts (e.g. England's southern coast) sand is washed away leaving behind a pebble beach.
- The diagram shows the profile of a typical sandy beach, including **sand dunes**.

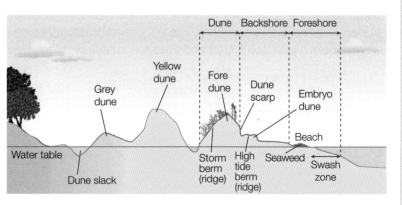

Figure 1 *Cross-section though beach and sand dunes*

Sand dunes

At the back of the beach, sand blown inland can build up to form *dunes*.

Embryo dunes form around obstacles (e.g. rocks).

↓

Dunes develop and are stabilised by vegetation (e.g. marram grass) to form fore dunes and tall yellow dunes.

↓

Decomposing vegetation makes sand more fertile and a wider range of plants colonise the back dunes.

↓

Ponds (dune slacks) can form in depressions.

Figure 2 *Development of sand dunes*

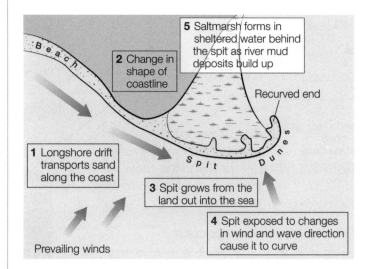

Figure 3 *Formation of a spit*

Spits and bars

- A **spit** is a long finger of sand or shingle jutting out into the sea.
- **Bars** form when longshore drift causes spits to grow across a bay.
- *Offshore bars* form further out to sea where waves approaching a gently sloping coast deposit sediment (due to friction with the sea bed).
- In the UK, some offshore bars have been driven on shore by rising sea levels.
- These are called *barrier beaches* – e.g. Chesil Beach (Dorset).

Six Second Summary

- Coastal deposition creates landforms such as beaches, sand dunes, spits and bars.

Over to you

Create a word cloud of words to do with coastal deposition processes and landforms.

You need to know:

- an example of a coastline in the UK
- how to identify if its landforms are caused by erosion and deposition.

*Student Book
See pages 102–3*

Example

Swanage, Dorset

Swanage lies on the south coast of England. The surrounding coastline has a range of coastal erosion and deposition landforms influenced by different rock types and geological structure. Rocks have been folded and tilted so that different rock types reach the coast.

Poole Harbour – one of the UK's largest natural harbours. Two spits have formed at the mouth.

Studland Bay – there are lagoons, saltmarshes, sand dunes and beaches

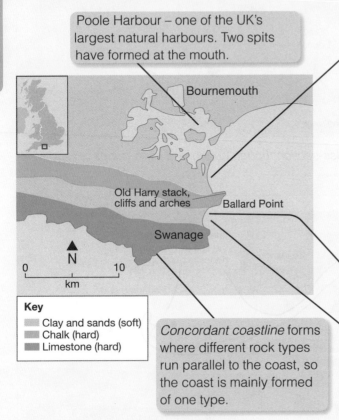

Bournemouth

Old Harry stack, cliffs and arches

Ballard Point

Swanage

0   N   10
    km

Key
- Clay and sands (soft)
- Chalk (hard)
- Limestone (hard)

Discordant coastline forms where there are alternating bands of harder (more resistant) and softer (less resistant) rocks. This creates headlands and bays.

Concordant coastline forms where different rock types run parallel to the coast, so the coast is mainly formed of one type.

Figure 1 *Geology of the Swanage coast*

Swanage Bay is sheltered with a broad, sandy beach

Add a WOW! factor

Swanage sits on the Jurassic Coast – so-called because it is important from a geological point of view, and Jurassic is the name of a geological period.

 Six Second Summary

Different rock types and geological structure are important factors in the formation of erosional and depositional landforms around the coast near Swanage.

 Over to you

Learn this example!

- Where is Swanage?
- What factors affect the formation of the features on this coastline?
- What are concordant/discordant coasts? Which of these applies to Swanage?
- What coastal features can you identifyand name?
- How have they formed?

You need to know:

- how to use map and photo evidence to identify coastal landforms of erosion and deposition.

Student Book
See pages 104–5

Using an OS map extract and photo

Figure **1** is an extract from an OS map of the Swanage coast. Figure **2** is an aerial view of the coast between Ballard Point and the Foreland. See Figure **1** opposite for the geology of the area.

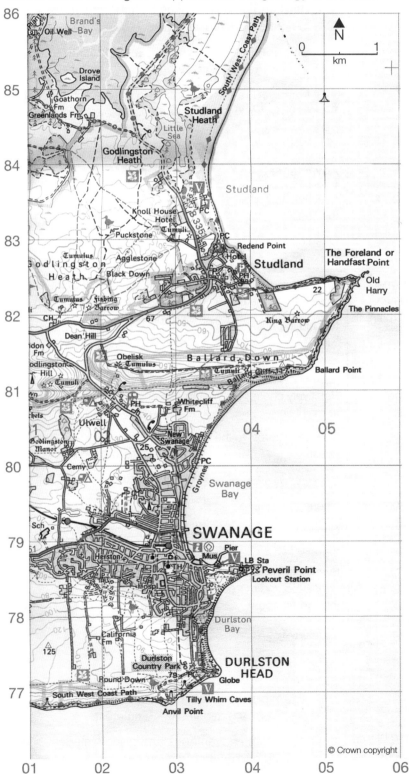

Figure 1 *1:50 000 OS map extract of Swanage coast*

Big Idea

You need to be able to identify these features:

headlands – The Foreland (Handfast Point), Ballard Point, Peveril Point, Durlston Head

bays – Studland Bay, Swanage Bay, Durlston Bay

beaches – in Studland Bay and Swanage Bay

stack – Old Harry.

Figure 2 *Aerial view of the coastline between Ballard Point and The Foreland*

Six Second Summary

OS maps can be used to identify landforms and to help interpret photos.

Over to you

Create a spider diagram with **one** leg for each different type of landform (headland etc.) you can identify on the OS map.

Include one example for each type of landform with a 6-figure grid reference.

Colour code the landforms as either:

- erosion features
- deposition features.

*Student Book
See pages 106-7*

You need to know:

- about the costs and benefits of hard engineering methods to protect the coast.

Managing coasts

Coasts are managed to protect people from erosion and flooding. But coastal defences are expensive, and in some cases increasing costs might outweigh the benefits. In future, some coastlines could be left undefended.

 **Big Idea**

The options for managing the coast are:

1 **Hard engineering** – uses artificial structures to control natural processes.

2 **Soft engineering** – involves methods that work with natural processes.

3 **Managed retreat** – controlled retreat of the coast (e.g. by allowing the sea to flood low-lying land).

Hard engineering

Method	Cost	Advantages	Disadvantages
Sea walls – concrete or rock barrier at the foot of cliffs or top of beach. Curved to reflect waves out to sea.	£5000–£10000 per metre	Effective at stopping the sea. Often creates a walkway.	Can look obtrusive and unnatural. Very expensive; high maintenance costs.
Groynes – rock or timber structures built at right angles to beach. They trap sediment moved by longshore drift and enlarge the beach. Wider beach reduces wave damage.	Timber groynes £150000 each (every 200 metres)	Create a wider beach – good for tourism. Not too expensive.	Interrupting longshore drift can lead to increased erosion elsewhere. Unnatural and rock groynes are unattractive.
Rock armour – piles of large boulders at foot of cliff. Rocks absorb wave energy to protect the cliff.	£200000 per 100 metres	Relatively cheap; easy to maintain. Can add interest to the coast.	Rocks are often from elsewhere and don't fit in with local geology. Expensive to transport rock. Can be obtrusive.
Gabions – rock-filled wire cages that support a cliff and provide a buffer against the sea.	Up to £50000 per 100 metres	Cheap to produce. Can improve cliff drainage. Eventually become vegetated and merge into landscape.	Unattractive initially. Cages rust within 5–10 years.

 Six Second Summary

- There are **three** options for protecting the coast from the sea.
- Hard engineering methods include the use of sea walls, groynes, rock armour and gabions.

 Over to you

Exam practice! This question is worth 6 marks – so you've 6 minutes to complete it.

Assess the advantages and disadvantages of hard engineering at the coast.

Student Book
See pages 108–9

You need to know:

- about the costs and benefits of soft engineering methods to protect the coast.

Soft engineering

Soft engineering schemes are generally cheaper than hard engineering, though they may need more maintenance (e.g. beaches need more sand/shingle every few years). But they are more sustainable and are the preferred option for coastal management.

Method	Cost	Advantages	Disadvantages
Beach nourishment – sand or shingle is dredged offshore and transported to the coast by barge. It is dumped on the beach and shaped by bulldozers creating a wider, higher beach (known as **re-profiling**). Beach protects land and property (Figure **1**).	Up to £500 000 per 100 metre	Blends in with existing beach. Bigger beach increases tourist potential.	Needs constant maintenance. Expensive
Dune regeneration – marram grass is planted to stabilise dunes and help them develop, which makes them effective buffers to the sea. Fences keep people off newly planted areas (Figure **2**).	£200–£2000 per 100 metre	Maintains a natural environment – good for wildlife. Relatively cheap.	Time-consuming to plant grass and construct fencing. Can be damaged by storms.
Dune fencing – fences are constructed along the seaward side of existing dunes to encourage new dune formation. New dunes help to protect existing dunes.	£400–£2000 per 100 metre	Little impact on natural systems. Controlling access protects other ecosystems.	Can be unsightly. Needs regular maintenance.

Figure 1 *Beach nourishment at Eastbourne, East Sussex*

Figure 2 *Dune regeneration at Chichester, West Sussex*

Six Second Summary

- Soft engineering is a more sustainable way of managing the coast as it works with natural processes.
- Soft engineering methods include beach nourishment and re-profiling, dune regeneration and dune fencing.

Over to you

Check you know what *sustainability* means.

Then explain why soft engineering methods are generally more sustainable than hard engineering methods as a way of protecting the coast.

*Student Book
See pages 110–11*

You need to know:

• how managed retreat and coastal realignment works.

Monitoring and adaptation

Where coastlines include low-value farmland, forest or moor, these coastal zones may not be protected – a 'Do nothing' approach. People living in these areas must **adapt** by moving further inland.

Scientists **monitor** these stretches of coast. This helps to reduce conflict and involves studying marine processes, mass movement and human activity to make sure 'Do nothing' is still appropriate.

Medmerry managed retreat

The flat, low-lying land at Medmerry, near Chichester in southern England, is mainly used for farming and caravan parks. In the past, it was protected by a low sea wall.

The land is of relatively low value, so the sea was allowed to breach the sea wall in 2013 and flood some of the farmland.

Costing £28 million, this managed retreat scheme will:

• create a large natural saltmarsh (a natural buffer to the sea)
• help to protect surrounding farmland and caravan parks from flooding
• establish a wildlife habitat and encourage visitors to the area.

Embankments have been built inland to protect farmland, roads and settlement. Alteration of the coastline like this is called *coastal realignment*.

 **Big Idea**

Managed retreat allows the sea to flood or erode an area of relatively low-value.

Figure 1 *Managed retreat – Medmerry, West Sussex*

 Six Second Summary

• Managed retreat is a form of soft engineering, which is used to manage the coast where land is of relatively low value.
• People living or working in these areas must adapt or move.

Over to you

1 List the opinions that any **six** different groups might have about managed retreat – e.g., farmers, caravan park owners.

2 Which of your groups might be in favour of managed retreat and who might be against?

3 Where might conflict arise?

Student Book
See pages 112–13

You need to know:

- why a coastal management scheme was needed at Lyme Regis
- the features of the scheme, and the outcomes.

Where is Lyme Regis?

Lyme Regis is a small coastal town in Dorset, on England's south coast, and is popular with tourists.

What are the issues?

- Unstable cliffs.
- Powerful waves from the south west cause rapid erosion.
- Foreshore erosion has destroyed or damaged many properties.
- Sea walls have been breached many times.

How has the coastline been managed?

The Lyme Regis Environmental Improvement Scheme was set up in the early 1990s to provide long-term coastal protection and reduce the threat of landslips. Engineering works were completed in 2015.

Key features of the scheme

Phases 1 and 2:

- new sea walls and promenades
- cliffs stabilised
- creation of wide sand and shingle beach to absorb wave energy
- extension of rock armour to absorb wave energy and retain beach.

Phase 4:

- new sea wall for extra protection
- cliffs stabilised to protect homes.

Total cost: over £43 million.

Remember: A planned **Phase 3** didn't go ahead – costs outweighed benefits.

How successful has it been?

Positive outcomes
<ul><li>New beaches have increased visitor numbers and seafront businesses are doing well.</li><li>New defences have withstood stormy winters.</li><li>Harbour is better protected.</li></ul>

Negative outcomes
<ul><li>Increased visitor numbers has caused conflict due to traffic congestion and litter.</li><li>Some think the new defences spoil the landscape.</li><li>The new sea wall might interfere with natural processes and cause problems elsewhere.</li></ul>

Big Idea

The Lyme Regis scheme used **both** hard and soft engineering.

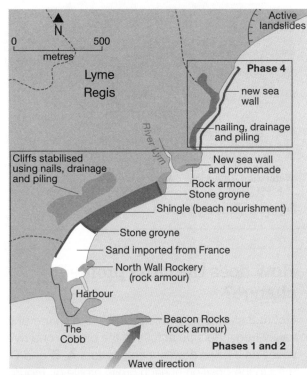

Figure 1 Coastal management at Lyme Regis

Six Second Summary

- Lyme Regis is built on cliffs that are being rapidly eroded.
- The management scheme aimed to protect the town and reduce the threat of landslips.
- The scheme was completed in phases over a period of more than 20 years.

Over to you

Learn this example! Make sure you know:

- *where* Lyme Regis is
- *what* the coastal management issues were
- *how* the coastline has been *managed* – note the *key features*
- *how successful* it has been.

Changes in rivers and their valleys

- what a drainage basin is
- how the long and cross profile of a river changes downstream.

Student Book
See pages 114–15

What is a drainage basin?

A *drainage basin* is the area of land drained by a river and its tributaries.

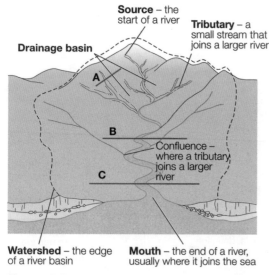

Figure 1 *Drainage basin*

How does a river's long profile change?

Figure **2** shows a river's **long profile** and how its gradient changes downstream. Put simply – it's steep in upland areas (the river's upper course), and gentle in the lowlands (the river's lower course).

In reality it varies, e.g. a waterfall creates a step in the long profile.

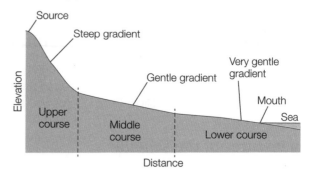

Figure 2 *Long profile of a river*

How does the cross profile change?

Figure **3** shows the **cross profile** – the shape of the valley from one side across to the other – of a river valley as it flows downstream. Letters **A**, **B** and **C** refer to the position of each cross profile on Figure **1**. Changes are due to the amount of water flowing in the river. As tributaries add more water (and energy) to the river, it erodes its channel, making it wider and deeper.

Changes to the valley cross profile are mainly due to channel erosion, broadening and flattening the base of the valley. Together with weathering and mass movement, these processes make the sides of the valley less steep.

Figure 3 *River and valley cross profiles*

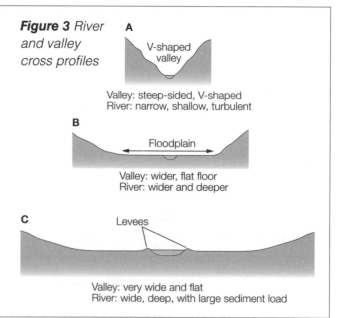

- A drainage basin is the area drained by a river and its tributaries.
- A long profile shows how the gradient of a river changes downstream.
- A cross profile is a cross-section of a river valley.

Make some 'key term' flashcards for: drainage basin, source, tributary, confluence, mouth, watershed, long profile, cross profile.

Now write a *geographical* definition for each term. Test yourself with a friend!

Student Book
See pages 116–17

You need to know:

- different processes of erosion
- how a river's load is transported
- when deposition happens.

What are the processes of erosion?

There are two types of erosion:

- **vertical** (downwards)
- **lateral** (sideways).

These combine to change the river channel and the river valley as the river flows downstream – see 11.1.

There are four processes of river erosion – see Figure **1**.

1 Hydraulic action	The force of water hitting the river bed and banks. Most effective when water is moving fast and at high volume.
2 Abrasion	The load carried by the river hits the bed or banks, dislodging particles.
3 Attrition	Stones carried by the river knock against each other, becoming smaller/more rounded.
4 Solution	Alkaline rocks, e.g. limestone, are dissolved by slightly acidic river water.

Figure 1 Processes of river erosion

What are the processes of transportation?

Material transported by a river is called its *load*. The four main types of river transportation are shown in Figure **2**.

The size and amount of load carried depends on a river's speed, or *velocity*. After heavy rain rivers look muddy – they are fast flowing and transporting large amounts of sediment. At low flow, rivers look clearer and little sediment is transported.

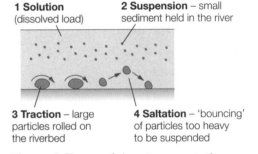

Figure 2 Types of river transportation

When does deposition happen?

When a river's velocity decreases, and it no longer has the energy to transport its load, it deposits it.

- Larger rocks transported mainly by **traction** are only carried short distances during periods of *high flow*. They are deposited in a river's upper course.
- Smaller sediment is carried further downstream – mostly in **suspension**. It is deposited on a river's bed and banks where velocity slows due to friction.
- Lots of deposition occurs at a river's mouth where its velocity reduces because of the gentle gradient and also by interaction with tides.

Figure **3** shows the balance between erosion, transportation and deposition along a river's course and the landforms that are created as a result.

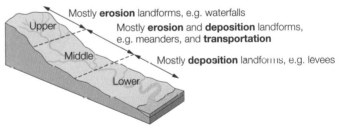

Figure 3 Processes and landforms along a river's course

Six Second Summary

- Vertical and lateral erosion combine to change a river's channel and valley downstream.
- There are **four** processes of river erosion and **four** types of river transportation.
- Deposition happens when a river's velocity decreases.

Over to you

Make up a mnemonic (a phrase or sentence, with words beginning with the first letter of each process) to help you remember the **four** processes of river erosion (HAAS), and **four** types of transportation (SSTS).

Student Book
See pages 118–19

- the characteristics and formation of interlocking spurs, waterfalls and gorges.

River erosion landforms

1

Interlocking spurs

In Figure **1**, a mountain stream erodes vertically creating a *V-shaped valley*. It winds around areas of resistant rock to create **interlocking spurs** which jut out into the valley.

V-shaped valley

Interlocking spurs

Figure 1 V-shaped valley and interlocking spurs

2

Waterfalls

As a river flows downstream it crosses different rock types. More resistant rocks are less easily eroded than less resistant rocks, forming steps in a river's long profile. The steps form **waterfalls** – see Figure **2**.

Waterfalls can also form:

- when sea level drops causing a river to cut down into its bed creating a step (called a *knick point*)
- in glacial **hanging valleys**.

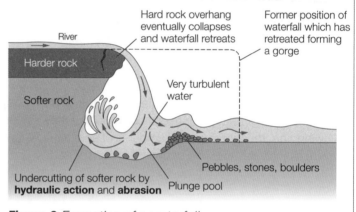

River

Harder rock

Softer rock

Hard rock overhang eventually collapses and waterfall retreats

Former position of waterfall which has retreated forming a gorge

Very turbulent water

Undercutting of softer rock by **hydraulic action** and **abrasion**

Plunge pool

Pebbles, stones, boulders

Figure 2 Formation of a waterfall

3

Gorges

A **gorge** is a narrow, steep-sided valley found downstream of a retreating waterfall (see Figure 2).

Gorges can form in other ways:

- at the end of the last glacial period masses of water from melting glaciers poured off upland areas forming gorges (e.g. Cheddar Gorge, Somerset)
- on limestone, when large underground caverns can accommodate an entire river.

 **Big Idea**

Erosion is the dominant process in a river's upper course creating **interlocking spurs, waterfalls** and **gorges**.

 Six Second Summary

- Each stage of a river's course (upper, middle, lower) has distinctive landforms.
- The main river erosion landforms in a river's upper course are interlocking spurs (in V-shaped valleys), waterfalls and gorges.

 Over to you

Draw a sketch of a river in its upper course to show its V-shaped valley and interlocking spurs. Then draw a diagram to show how waterfalls and gorges form.

Add annotations to explain how the landforms are created.

River erosion and deposition landforms

11.4

You need to know:

- the characteristics and formation of meanders, ox-bow lakes, levees, floodplains and estuaries.

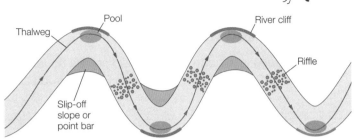

Student Book
See pages 120–1

River erosion *and* deposition landforms

Meanders

- **Meanders** are bends in a river found mainly in lowland areas. They constantly change shape and position.
- Figure **1** shows the main features and processes taking place in a meandering river. The *thalweg* is the line of fastest current. It swings from side to side causing erosion on the outside bend, and deposition on the inside bend. These processes cause meanders to migrate across the valley floor.

Ox-bow lakes

As meanders migrate across the valley floor they erode towards each other, eventually forming an **ox-bow lake**.

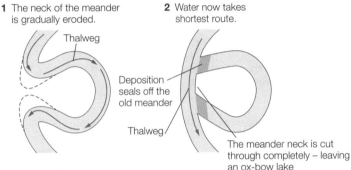

Figure 1 *Meandering river – processes and landforms*

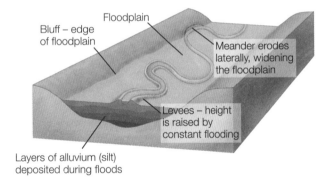

Figure 2 *Formation of an ox-bow lake*

River deposition landforms

Floodplains and levees

Floodplains are wide, flat areas on either side of a river in its middle and lower courses. They are created by migrating meanders and floods depositing layers of silt to form alluvium.

Levees form when, in low flow, deposition raises the river bed so the channel can't carry as much water. During flooding, water flows over the sides of the channel. As velocity decreases, coarser sediment is deposited first on the banks – then finer sand and mud, raising the height of the levees.

Estuaries

An **estuary** is where the river meets the sea. They are affected by tides, wave action *and* river processes. As the tide rises, rivers can't flow into the sea, so velocity falls and sediment is deposited forming *mudflats*, which develop into *salt marshes*.

Figure 3 *Formation of floodplains and levees*

Six Second Summary

- Meanders and ox-bow lakes form as a result of erosion *and* deposition.
- Floodplains and levees are formed by deposition.
- The main process operating in estuaries is deposition.

Over to you

Outline the processes that create the following landforms: meanders, ox-bow lakes, floodplains, levees. Draw labelled diagrams to explain the formation of **two** of these.

Student Book
See pages 122–3

Example

You need to know:

- an example of a river valley in the UK
- how to identify, describe and explain its major landforms of erosion and deposition.

Where is the River Tees?

The River Tees is in north-east England and its source is in the Pennine Hills. It flows roughly east to reach the North Sea at Middlesbrough.

Landforms of erosion

High Force (Figure **1**) on the River Tees is in the river's upper course. The waterfall drops 20 m and continues through a gorge.

A resistant band of igneous rock (dolerite) cuts across the valley. Its resistance has led to the development of a waterfall.

Underlying weaker rock (limestone) is undercut.

Waterfall retreats upstream to form a gorge.

Figure 1 *High Force on the River Tees*

Landforms of deposition

The map shows the River Tees south of Darlington. The river is flowing west to east over relatively flat, low-lying land. Along this stretch of the river are examples of meanders, levees, and floodplains.

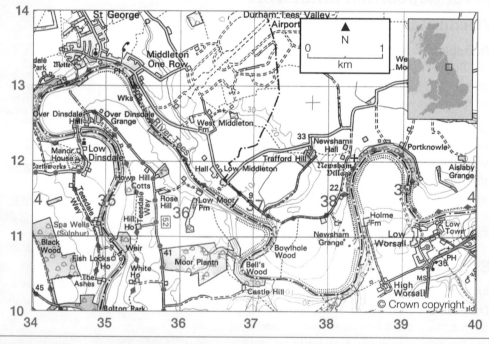

Figure 2 *1:50 000 OS map extract of River Tees, near Darlington*

 Six Second Summary

The River Tees has good examples of erosion landforms (High Force waterfall and gorge) and deposition landforms (levées and floodplains).

 Over to you

Look at Figure **2** and give a four-figure grid reference for:

- a meander (large bend in the river)
- levees (shown by this symbol """" """")
- a floodplain (white areas alongside river with few contour lines).

Factors increasing flood risk

Student Book
See pages 124–5

You need to know:

- what factors affect the risk of flooding
- what a hydrograph is, and what affects its shape.

What causes flooding?

- River floods usually occur after long periods of rain – most frequently during winter.
- Sudden floods, called *flash floods*, tend to occur in summer and are associated with heavy rainstorms.
- Physical and human factors can each increase flood risk.

 **Big Idea**

A river flood occurs when a river channel can no longer hold the amount of water flowing in it. Water overspills the banks onto the floodplain.

Physical factors	Human factors (land use)
• *Precipitation* – torrential rainstorms and/or prolonged periods of rain can lead to flooding. • *Geology* – impermeable rocks don't allow water to pass through, so it flows overland into river channels. • *Relief* – steep slopes mean water flows quickly into river channels.	• *Urbanisation* – impermeable surfaces, e.g. tarmac roads, mean water flows quickly into drains, sewers and river channels. • *Deforestation* – when trees are removed, much of the water which had been evaporated from leaves or stored on leaves and branches flows rapidly into river channels. • *Agriculture* – exposed soil can lead to increased surface runoff (especially if ploughing occurs up and down slopes).

What is a hydrograph?

A hydrograph shows how a river reacts to a rainfall event. It shows rainfall and **discharge** – the volume of water flowing in a river measured in m³ per second (cumecs). *Lag time* shows how quickly water is transferred into the river channel. The shorter the lag time, the greater the risk of flooding.

These factors affect its shape:

- basin size
- drainage density
- rock type and permeability
- land use
- relief
- soil moisture
- rainfall intensity
- antecedent rainfall

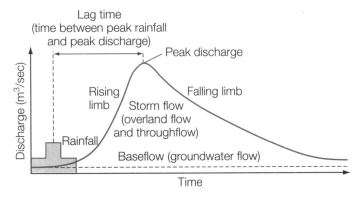

Figure 1 *A flood hydrograph*

Six Second Summary

- Flooding occurs when a river can't hold the amount of water flowing in it.
- Human and physical factors increase the flood risk.
- A flood hydrograph shows how a river reacts to a rainfall event.
- Different factors affect the shape of a hydrograph.

Over to you

Draw a larger version of the two hydrographs below. Annotate each one to explain what factors have affected their shape.

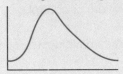

1 'Flashy' hydrograph with a short lag time and high peak

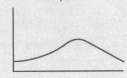

2 A flat hydrograph with a low peak

Student Book
See pages 126–7

You need to know:

- what hard engineering is, and some examples of it
- about costs and benefits of hard engineering strategies to manage river flooding.

What is hard engineering?

Hard engineering involves using artificial structures to prevent, or control flooding. It is usually very expensive and the *costs* have to be weighed against *benefits*.

- Costs include financial costs and negative impacts on people/the environment.
- Benefits are the financial savings and environmental improvements made by preventing flooding.

Dams and reservoirs	Channel straightening	Embankments	Flood relief channels
• Widely used to regulate river flow and reduce risk of flooding. • Often multi-purpose, e.g. flood prevention; HEP generation; water supply. • Can be effective in regulating water flow and can store water in reservoir. • Expensive, and reservoirs often flood large areas of land.	• Cutting through meanders creating a straight channel, speeding up water flow. But can increase flood risk downstream. • Straightened channels may be lined with concrete. This can be unattractive and can damage wildlife habitats.	• Raise the level of a river bank allowing the channel to hold more water to help prevent flooding. • Concrete or stone walls are often used in towns, though mud dredged from the river can be used. This is cheaper, more sustainable and looks more natural.	• These can be built to by-pass urban areas. At times of high flow, sluice gates allow excess water to flow into the flood relief channel, reducing the threat of flooding.

Figure 1 *Hard engineering strategies*

Clywedog reservoir, Llanidloes

The Clywedog reservoir was built in the 1960s to help prevent flooding of the River Severn. The reservoir stretches for nearly 10km. It fills in the winter and water is released in the summer to maintain a constant flow.

Figure 2 *Clywedog dam and reservoir*

Jubilee River, Maidenhead

The Jubilee River is an 11km long flood relief channel built to reduce the flood risk on the Thames. It opened in 2002. It has had a positive impact on the environment by creating new wetlands.

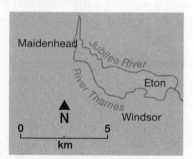

Figure 3 *The Jubilee River*

  Six Second Summary

- Hard engineering involves the use of artificial structures to prevent or control flooding.
- Hard engineering schemes have costs and benefits.

Over to you

Use two highlighters to highlight the *costs* and *benefits* of strategies listed in Figure **1**. Summarise each strategy under the following headings:

- What is it?
- How does it work?
- Advantages
- Disadvantages

Student Book
See pages 128–9

You need to know:

- what soft engineering is, and some examples of it
- the costs and benefits of soft engineering strategies to manage river flooding
- about preparing for floods.

What is soft engineering?

Soft engineering involves working *with* natural processes to manage flood risk. It aims to reduce and slow movement of water into a river channel to help prevent flooding. As with hard engineering, there are costs and benefits.

Afforestation (planting trees)	Wetlands and flood storage areas	Floodplain zoning	River restoration
• Trees obstruct the flow of water, and slow down its transfer to river channels. • Water is taken up by trees and evaporated from leaves and branches. • It is relatively cheap with environmental benefits.	• Wetlands are deliberately allowed to flood, forming storage areas. • Reduces the risk of flooding downstream.	• Restricts different land uses to certain zones on the floodplain. • Areas at risk from flooding can be used for grazing, parks and playing fields. • Can reduce losses caused by flood damage. • Can be difficult to implement on already developed land.	• When a river's course has been changed artificially, it can be restored to its original course. • It uses the natural processes and features of a river, e.g. meanders and wetlands to slow down flow and reduce the likelihood of flooding downstream.

Figure 1 *Soft engineering strategies*

Preparing for floods

In England and Wales, the Environment Agency issues **flood warnings**. There are three levels:

- *Flood watch* – flooding of low-lying land and roads expected. Be prepared.
- *Flood warning* – a threat to homes and businesses. People should move valuable items upstairs and turn off electricity and water.
- *Severe flood warning* – extreme danger to life and property. People should stay upstairs or leave their home.

The Environment Agency produces flood maps showing areas at risk of flooding. People living in these areas should plan for floods by using sandbags and floodgates to prevent water damaging property.

Local authorities and emergency services use flood maps to plan their responses to floods including installing temporary flood barriers, evacuating people and closing roads.

Six Second Summary

- Soft engineering works with natural processes to manage flood risk, and includes the use of afforestation, wetlands and flood storage areas, floodplain zoning and river restoration.
- The Environment Agency produces flood maps and issues flood warnings.

Over to you

Use two highlighters to highlight the *costs* and *benefits* of strategies listed in Figure **1**. Summarise each strategy under the following headings:

- What is it?
- How does it work?
- Advantages
- Disadvantages

How would you decide whether hard or soft engineering is better if asked in an exam? Give a precise answer – no waffle.

Managing floods at Banbury

Managing floods – soft engineering

Example

Student Book
See pages 130–1

You need to know:

- why a flood management scheme was needed in Banbury
- what the scheme consists of
- the costs and benefits of the scheme.

Where is Banbury?

- Banbury is about 30 km north of Oxford.
- The population is about 45 000.
- Much of the town is on the floodplain of the River Cherwell (a tributary of the Thames).

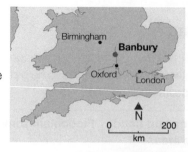

Figure 1 *Location of Banbury*

Add a WOW! factor

A 200-year flood event is a flood which is expected to happen once every 200 years.

Why was the scheme needed?

Banbury has a history of flooding.

- In 1998 flooding closed the railway station, shut roads and caused £12.5 million of damage.
- In 2007 it was flooded again (along with much of central and western England).

What has been done?

In 2012 the flood defence scheme was completed. Figure **2** shows what was done.

Social	• The raised A361 stays open during a flood avoiding disruption. • Quality of life improved with new footpaths and green areas. • Less anxiety about flooding.
Economic	• The scheme cost £18.5 million, paid for partly by the Environment Agency and Cherwell District Council. • Over 400 houses and 70 businesses protected at a value of over £100 million.
Environmental	• Earth needed to build embankment was extracted locally, creating a small reservoir. • A new habitat has been created with ponds, trees and hedges.

Figure 3 *Costs and benefits*

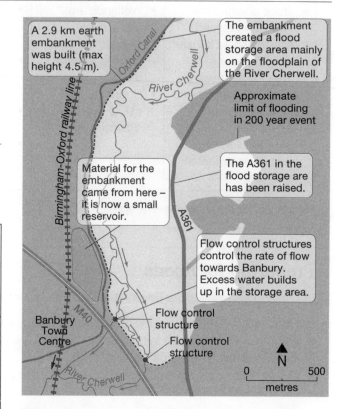

Figure 2 *Features of the Banbury Flood Storage Reservoir. A new pumping station transfers excess water into the river below the town.*

Six Second Summary

- Banbury is on the floodplain of the River Cherwell and at risk from flooding.
- A flood defence scheme has created a flood storage area and allowed the flow of the river to be controlled.
- The scheme has costs and benefits.

Over to you

Create a poster, flashcards, spider diagram to help you learn this example. You need to know:

- *where* is Banbury?
- *why* the scheme was needed
- *what* the flood defence scheme consists of (what has been done)
- the social, economic and environmental *issues*.

Student Book
See pages 132–3

You need to know:

- how far ice extended across the UK during the last ice age
- the processes operating in glacial environments.

A glacial environment

During the last ice age, snow and ice covered much of the UK. Glaciers in the north and west carved deep valleys and troughs. Further south and east land was permanently frozen.

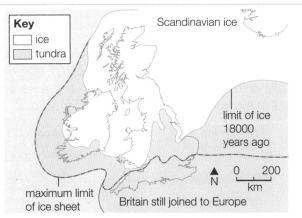

Figure 1 Ice across the UK during the last ice age

Weathering processes

The main process in glacial environments is **freeze-thaw** (see 12.1). It's mainly seasonal – water freezes in winter, and thaws in summer. Freeze-thaw:

- helps create a jagged landscape of frost-shattered rock
- weakens rocks so they are more easily eroded
- creates *scree* which acts as a powerful erosion tool when trapped under moving glaciers.

Erosion processes

There are two main types of glacial erosion:

- **abrasion** – a 'sandpaper' effect caused by ice scouring the valley floor.
- *Striations* (scratches) are caused by large rocks below the ice.
- **plucking** occurs when meltwater beneath a glacier freezes around rock. Loose rock is 'plucked' away as the glacier moves over it.

Glacial movement

- In summer, meltwater lubricates the glacier so that it can slide downhill – *basal slip*. In hollows high on the valley side, the movement may be more curved – **rotational slip**.
- In winter the glacier is frozen to the rock surface. The weight of the ice and effect of gravity cause ice crystals to change shape. This is called *internal deformation* and causes the glacier to move slowly downhill

Glacial transportation

- Sediment carried by a glacier – called **moraine** – can be transported *on, in* or *below* the ice.
- As a glacier moves it pushes loose material ahead of it – it's called **bulldozing**.

Glacial deposition

- Deposition occurs when ice melts; most occurs at the glacier's *snout* (the front).
- As a glacier melts and retreats, it leaves behind poorly sorted rock fragments called **till** or *boulder clay*.
- In front of the glacier, meltwater transports sediment away. Larger rocks are deposited close to the ice, finer material is carried further away. This sandy and gravel material is called **outwash**.

Six Second Summary

- During the last Ice Age much of the UK was covered by ice.
- Processes operating in glacial environments include weathering, erosion, movement, transport and deposition.

Over to you

Create a flow diagram to show how the processes of weathering, erosion, glacial movement, transport and deposition are linked in a glacial environment.

You need to know:

- the characteristics and formation of landforms resulting from glacial erosion.

Student Book
See pages 134–5

Corries

Also know as **cirques** and **cwms** – **corries** are large depressions found on the upper slopes of glaciated valleys. They have a steep back wall and a raised 'lip' at the front. Figure **1** shows their formation.

 **Big Idea**

Ice is a powerful agent of erosion, creating spectacular landforms in mountainous areas.

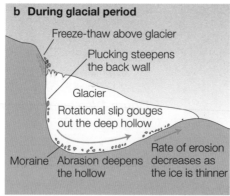
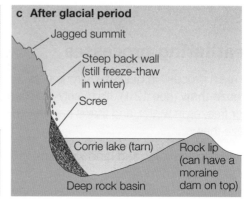

a Beginning of glacial period
Snow accumulates in hollow; compressed into ice

b During glacial period
Freeze-thaw above glacier
Plucking steepens the back wall
Glacier
Rotational slip gouges out the deep hollow
Moraine Abrasion deepens the hollow
Rate of erosion decreases as the ice is thinner

c After glacial period
Jagged summit
Steep back wall (still freeze-thaw in winter)
Scree
Corrie lake (tarn)
Rock lip (can have a moraine dam on top)
Deep rock basin

Figure 1 *Formation of a corrie*

Arêtes and pyramidal peaks

- An **arête** is a narrow knife-edged ridge separating two corries.
- Arêtes typically form when erosion occurs in two adjacent corries.
- If three or more corries erode back-to-back a **pyramidal peak** may form.

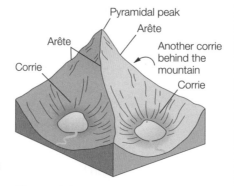

Pyramidal peak
Arête
Arête
Corrie
Another corrie behind the mountain
Corrie

Figure 2 *Arêtes and pyramidal peaks*

 Six Second Summary

- Ice is a powerful agent of erosion.
- Corries, arêtes and pyramidal peaks are found on the upland parts of glaciated valleys.
- Glacial valley landforms include glacial troughs, truncated spurs, hanging valleys and ribbon lakes.

Glacial valley landforms

Most glaciers flow along already existing river valleys. They can't flow round obstacles, so carve straight courses.

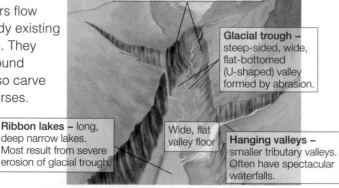

Truncated spurs – glaciers cut straight through interlocking spurs (see 11.3)

Glacial trough – steep-sided, wide, flat-bottomed (U-shaped) valley formed by abrasion.

Ribbon lakes – long, deep narrow lakes. Most result from severe erosion of glacial trough.

Wide, flat valley floor

Hanging valleys – smaller tributary valleys. Often have spectacular waterfalls.

Figure 3 *Glacial valley landforms*

Over to you

What's the difference between an arête and a pyramidal peak; a glacial trough and a hanging valley; a ribbon lake and a corrie lake (tarn)?

Now, draw an annotated diagram to show the formation of **one** glacial erosion landform.

You need to know:

- the characteristics and formation of landforms resulting from glacial transport and deposition.

*Student Book
See pages 136–7*

Moraine

Glaciers act like conveyor belts carrying weathered and eroded rock – **moraine** – from the mountains to the lowlands (see 12.1 on transport and deposition). There are several types of moraine (Figure **1**).

As ice melts many of these features are eroded by meltwater.

Lateral moraine – forms at the edges of the glacier. Mostly consists of scree resulting from freeze-thaw weathering. When ice melts, it forms low ridges on the valley sides.

Medial moraine – forms when a tributary glacier joins the main glacier and two lateral moraines merge. On melting, medial moraine forms a ridge down the centre of the valley.

Freeze-thaw on valley sides

Scree

Ground moraine – material transported below a glacier and left behind when it melts. Often forms uneven hilly ground.

Terminal moraine – forms when material piles up at the glacier's snout, and forms a ridge across the valley. Represents the furthest extent of glacier's advance.

Figure 1 *Types of moraine*

Drumlins

- **Drumlins** are smooth, egg-shaped hills several hundred metres long. They are found in clusters on the floor of a glacial trough.
- They consist of moraine that has been shaped by the moving ice.
- They usually have a blunt end (facing up-valley) and a more pointed end (facing down-valley).

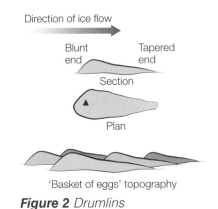

Figure 2 *Drumlins*

Erratics

- An **erratic** is a large boulder resting on a different type of rock.
- By studying the geology of an area it is possible to work out where an erratic came from.

Figure 3 *Erratics in the Scottish Highlands*

Six Second Summary

- There are **four** types of moraine: ground, lateral, medial and terminal.
- Drumlins are smooth, egg-shaped hills found on the floor of a glacial trough.
- Erratics are large boulders that have been transported by glaciers.

Over to you

What's the difference between lateral and medial moraine; ground and terminal moraine; a drumlin and an erratic?

Now, draw a labelled diagram to show how drumlins form.

Student Book
See pages 138–9

Example

You need to know:

- an example of an upland glaciated area in the UK
- how to identify its major landforms of erosion and deposition.

Identifying glacial landforms on an OS map

Cadair Idris is in Snowdonia National Park, north Wales, and has many glacial features. You can identify these on a 1:50 000 OS map (Figure **1**).

Erosional landforms

A **corrie** is shown as a series of semi-circular contours. The steep sides are shown by the pattern of bold black lines (cliff symbol).

The edges of the corrie form **arêtes**, shown by the black cliff symbol. The arête at the back of the Llyn Cau corrie is called Craig Cau.

You can see a **pyramidal peak** at spot height 791 where there are three corries back-to-back.

Tal-y-llyn Lake lies in a wide flat-bottomed, steep-sided **glacial trough**.

Note the shape of the contours on the **truncated spur**. They change from semi-circular at the top of the slope to straighter at the bottom. This shows where the old interlocking spur has been cut off.

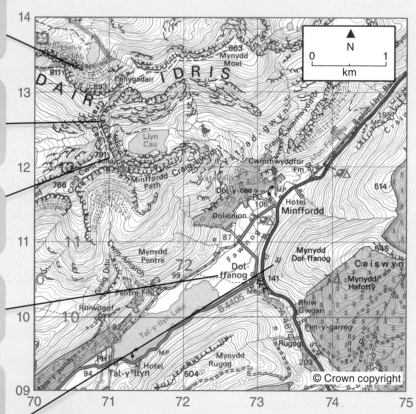

Figure 1 *A 1:50 000 OS map extract of Cadair Idris*

Depositional landforms

Large-scale features of glacial deposition are usually found in lowland areas, so there are few in mountainous landscapes, such as Cadair Idris.

But note that the black dots on the map in several of the corries are rock debris, which could be glacial deposits. Some moraine ridges are found in this area.

Six Second Summary

Cadair Idris is a mountainous landscape with distinctive glacial features and landforms which can be identified on OS maps.

Over to you

Match the following landforms with the correct grid reference:

corrie, ribbon lake, corrie lake (tarn), truncated spur, pyramidal peak, glacial trough, arête.

a 720100 **b** 727107 **c** 710118 **d** 715124 **e** 733110
f 709121 **g** 710125

Student Book
See pages 140–1

You need to know:

- how glaciated areas provide opportunities for farming, forestry, quarrying and tourism.

Farming

- In upland areas, glacial erosion stripped away soil and vegetation. Soils are thin and acidic and mainly used for grazing. Sheep can tolerate the cold, wet and wind, and the poor vegetation.

- Soils in valleys are thicker because of deposition. It's easier to use machinery on flat-bottomed glacial troughs, so crops including cereals and potatoes are grown. Land is also used for growing winter feed (hay and silage) for animals.

- Lowland glaciated areas such as eastern Britain may be covered by a thick layer of fertile till. This makes productive farmland, and wheat, barley, potatoes, etc. are grown.

Forestry

- Coniferous trees are adapted to cope with the acidic soils in upland glaciated areas of the UK and large plantations (mostly of conifers) have been planted across Scotland and parts of northern England.

- Conifers can be left to grow for 20–30 years before being cut down to produce 'soft' wood which is used for timber in the construction industry, or for making paper.

Quarrying

- Upland glaciated areas consist of hard, resistant rock which can be quarried and crushed for use in the construction industry and for road building.

- Limestone (found in the Pennine Hills) is used in the chemical industry, for improving soils and for making cement.

- In lowland areas, material deposited by meltwater streams is used – sand for making cement; gravel to make concrete.

Tourism

- The UK's upland glaciated areas attract tourists who enjoy outdoor activities and cultural heritage.

- Tourism provides employment for thousands of people.

- Aviemore (near the Cairngorm Mountains, Scotland) is one of the UK's main mountain activity centres with mountain biking, skiing, walking, climbing, and lots of wildlife.

 Six Second Summary

- Upland and lowland glaciated areas of the UK provide opportunities for economic activities including farming, forestry, quarrying and tourism.

Over to you

Create a spider diagram of economic opportunities in glaciated areas. Add a leg for each of: farming, forestry, quarrying, and tourism. Add as much detail to it as you can.

Now, using two highlighters, highlight the activities that occur in upland areas, then in a second colour, those that occur in lowland areas.

Student Book See pages 142–3

You need to know:

- about conflicts between different land uses, and between development and conservation, in glaciated areas.

Land use: wind farms

Kirkstone Pass is one of the Lake District's most remote valleys. A project to build three 16 m wind turbines was completed in 2012, and cost £150 000.

The turbines provide power for a pub which had relied on generators for heat and light. There was opposition, but the Friends of the Lake District supported the scheme. The group said that 'green power' was good for the environment, and helped to secure the future of the pub and its employees.

But arguments against wind farms elsewhere in the Lake District include:

- people think they spoil the natural landscape
- fewer tourists stay in the area, affecting the local economy
- house prices might fall if views are spoilt by turbines.

Big Idea

Development in glaciated areas can lead to **land use conflict**:

- *quarrying* can lead to land and river pollution and spoil the landscape
- *tourism* creates problems over access to land, traffic congestion and rising house prices
- building *reservoirs* can create environmental issues.

Development v. conservation: Glenridding zip-wire

In 2014 Treetop Trek, a Windermere-based company, put forward a proposal to construct four parallel one-mile long zip-wires above Glenridding in Patterdale, in the Lake District.

People had clear views on the proposal, and there was strong local opposition.

Figure 1 *The site and route of the proposed zip-wire, Glenridding, Patterdale*

The leader of the campaign to stop the scheme said:

'It was an example of the conflict of interest between the aims of conserving the natural beauty and heritage of the Lake District and becoming more commercial. Local people thought it would drive away visitors who came to enjoy peace and quiet.'

The view of Treetop Trek was that:

'Our priority is to balance the need to conserve the landscape with enhancing the economy and improving opportunities for visitors.'

The scheme didn't go ahead.

 Six Second Summary

- Opportunities for development in glaciated areas can lead to conflict over use of land and resources.
- Some people are opposed to the construction of wind farms in the Lake District.
- The proposed zip-wire is a good example of conflict between development and conservation.

 Over to you

Create a mind map of different land *uses* and *users* in the Lake District.

Add notes to the links to show where conflict might occur.

Student Book
See pages 144–5

You need to know:

- an example of a glaciated upland area in the UK, and about the attractions for tourists
- the impacts of tourism
- the strategies used to manage the impacts.

What are the attractions of the Lake District?

The Lake District in north-west England is an upland glaciated area and a National Park.

Physical attractions	Cultural attractions
• Lakes (e.g. Windermere) provide water sports, cruises and fishing. • Mountains (e.g. Helvellyn) are popular for walking and mountain biking. • Adventure activities include abseiling and rock climbing.	• Landscape has inspired poets (e.g. Wordsworth) and writers (e.g. Beatrix Potter, whose home is a tourist attraction). • Scenic towns and villages (e.g. Ambleside) are popular. • Monuments such as Muncaster Castle at Ravenglass.

Add a *WOW!* factor

Scafell Pike is England's highest mountain (977 m). Windermere is the UK's largest natural lake (17 km long)!

What are the impacts of tourism on the Lake District?

Social	Economic	Environmental
• 14.8 million visitors in 2014 (there are 40 000 residents!). • 90% of visitors arrive by car – causing congestion. • High house prices – 20% are holiday lets or second homes. • Most tourism jobs are seasonal and poorly paid.	• Tourists spent £1000 million in 2014 supporting hotels, shops, restaurants. • Provides thousands of jobs. • Congestion slows down business communications.	• Honeypot sites are overcrowded and footpaths are damaged. • Pollution from cars and boats damages ecosystems. • Walkers damage farmland and dogs disturb livestock.

Impact management strategies

Traffic congestion

- Building dual-carriageways to improve access.
- Creating transport hubs (e.g. at Ambleside), linking parking, buses, ferries, walking and cycling.
- Expanding park-and-ride bus schemes.
- Introducing traffic calming measures (e.g. speed bumps) in villages.

Footpath erosion

- The Upland Path Landscape Restoration Project and Fix the Fells repair paths and re-plant native plants.
- There are still hundreds of kilometres of footpaths that need on-going maintenance.

 **Over to you**

Create a poster, flashcards, spider diagram or whatever works best for you. You need to know:

- what *attracts* tourists to the Lake District
- the *impacts* of tourism
- what *strategies* have been used to manage the impacts.

 **Six Second Summary**

- The Lake District has many attractions for visitors.
- Tourism has a range of impacts on the area.
- Different strategies are used to manage tourism.

Section A
Urban issues and challenges

Your exam

Section A Urban issues and challenges makes up part of Paper 2: Challenges in the human environment.

Paper 2 is a one-and-a-half hour written exam and makes up 35 per cent of your GCSE. The whole paper carries 88 marks (including 3 marks for SPaG) – questions on Section A will carry 33 marks.

You need to study all the topics in Section A – in your final exam you will have to answer questions on all of them.

Tick these boxes to build a record of your revision

Your revision checklist

Spec key idea	Theme	1	2	3
13 The urban world				
A growing percentage of the world's population lives in urban areas	13.1 An increasingly urban world			
	13.2 The emergence of megacities			
	13.3 Introducing Rio de Janeiro			
Urban growth creates opportunities and challenges for cities in lower income countries and newly emerging economies	13.4 Social challenges in Rio			
	13.5 Economic challenges in Rio			
	13.6 Improving Rio's environment			
	13.7 Managing the growth of squatter settlements			
	13.8 Planning for Rio's urban poor			
14 Urban change in the UK				
Urban change in cities in the UK leads to a variety of social, economic and environmental opportunities and challenges	14.1 Where do people live in the UK?			
	14.2 Introducing Bristol			
	14.3 How can urban change create social opportunities?			
	14.4 How can urban change create economic opportunities?			
	14.5 How can urban change affect the environment?			
	14.6 Environmental challenges in Bristol			
	14.7 Creating a clean environment in Bristol			
	14.8 Social inequality in Bristol			
	14.9 New housing for Bristol			
	14.10 The Temple Quarter regeneration (1)			
	14.11 The Temple Quarter regeneration (2)			
15 Sustainable urban development				
Urban sustainability requires management of resources and transport	15.1 Planning for urban sustainability			
	15.2 Sustainable living in Freiburg			
	15.3 Sustainable traffic management strategies			

An increasingly urban world

You need to know:

* how the world's cities are growing.

*Student Book
See pages 148–9*

How is the world's population changing?

The world's population is increasing. Since the 1900s, the bigger the global population has become, the faster it has grown (Figure **1**).

Figure 1 *Global population growth since the year 1000*

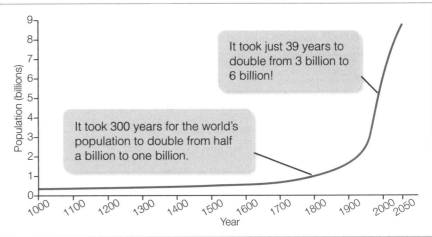

It took just 39 years to double from 3 billion to 6 billion!

It took 300 years for the world's population to double from half a billion to one billion.

What is urbanisation?

Urbanisation is the proportion of the world's population who live in cities.
Urbanisation is growing because of natural increase (births minus deaths) and migration.

Urban growth is the increase in the area covered by cities.

How does urbanisation vary around the world?

Urban population is growing more quickly in less developed regions than in more developed regions.

The largest growth in urban population by 2050 will take place in India, China and Nigeria. These are all examples of **newly emerging economies (NEEs)**.

In south and south-east Asia, around half the population live in towns and cities.

% population urban
- 75–100
- 50–74
- 25–49
- 0–24

In most **high-income countries (HICs)**, over 60% of the population live in cities.

In many **low-income countries (LICs)**, more than 20% live in cities. In Africa, the average urban population is almost 40%.

Figure 2 *Global urban population, 2014*

 Six Second Summary

* The global population has grown rapidly since 1950.
* Urbanisation is growing around the world.
* Currently, urbanisation is higher in HICs than in LICs.
* Urban populations are growing more quickly in less developed regions.

 Over to you

Write a clear definition of urbanisation. Write **five** statements that show how urbanisation varies in different parts of the world.

Student Book
See pages 150–1

You need to know:

- the factors that make cities grow.

Why do cities grow?

There are two main reasons why cities are getting bigger: natural increase and rural–urban migration.

Natural increase

Natural increase is where the birth rate is higher than the death rate.

Natural increase is higher in LICs and some NEEs because:

- there are lots of young adults aged 18–35
- improvements to health care have significantly lowered the death rate.

What are megacities?

Megacities are cities with a population of over 10 million.

There are three types of megacities:

Type	Features	Examples
Slow-growing	No squatter settlements	Tokyo Los Angeles (often in HICs)
Growing	Under 20% in squatter settlements	Beijing Rio de Janeiro (often in NEEs)
Rapid-growing	Over 20% in squatter settlements	Jakarta Mumbai (often in NEEs or LICs)

Figure 1 *Satellite image of Tokyo*

Rural–urban migration

Rural–urban migration is the movement of people from the countryside into towns and cities

This is caused by *push* and *pull factors*.

Push factors
Reasons why people want to leave the countryside.

- Farming is hard and poorly paid.
- Farming is often at subsistence level, leaving nothing to sell.
- Rural areas are isolated, often with few services.

Pull factors
Reasons why people are attracted to the city.

- A higher standard of living is possible.
- There are better medical facilities.
- There is a better chance of getting an education.

Six Second Summary

- Cities are growing because of natural increase and rural–urban migration.
- Rural–urban migration is caused by push factors and pull factors.
- Megacities have a population of over 10 million.
- Megacities are growing more quickly in NEEs and LICs.

Over to you

Make a list of **three** push factors and **three** pull factors. Make sure they're not just opposites of each other.

Introducing Rio de Janeiro

Student Book
See pages 152–3

You need to know:

- why the city of Rio de Janeiro is growing so rapidly.

Where is Rio?

Rio de Janeiro is situated in south-east Brazil, around Guanabara Bay.

It has four main zones:

Big Idea

Rio de Janeiro is major city in an NEE. It is a case study to show how urban growth creates *opportunities* and *challenges* for cities in LICs and NEEs.

North Zone – Industry, squatter settlements and the international airport

Centro – Historic buildings, CBD and financial centre

West Zone – Wealthy suburbs, industrial areas and Olympic stadiums

South Zone – Hotels, beaches and luxury flats, as well as Rocinha – the largest favela in South America

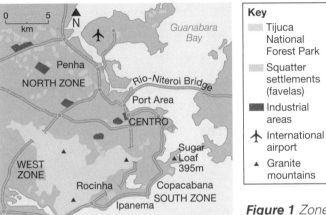

Key
- Tijuca National Forest Park
- Squatter settlements (favelas)
- Industrial areas
- ✈ International airport
- ▲ Granite mountains

Figure 1 Zones in Rio de Janeiro

Why is Rio an important city?

- The cultural capital of Brazil.
- A UNESCO World Heritage Site.
- Stunning natural surroundings.
- Host for 2016 Olympics and matches during the 2014 World Cup.
- Manufacturing industries e.g. chemicals and furniture.
- 'Christ the Redeemer' – one of the 'New' Seven Wonders of the World.
- A major regional, national and international industrial centre.
- A major port.
- Service industries, e.g. banking and finance.
- An important international transport hub.

Figure 1 Sugar Loaf Mountain, Rio

Why has Rio de Janeiro grown?

Economic activities attract migrants from many different places, including:

- the Amazon Basin
- Argentina and Bolivia
- South Korea and China seeking new business opportunities
- Portugal because of the common language (Portugal is Brazil's former colonial power)
- skilled workers from the USA and UK.

Six Second Summary

- Rio de Janeiro is in south-east Brazil.
- It is a major regional, national and international city.
- It has grown mainly because of migration, with people attracted to Rio's employment.

Over to you

Draw a spider diagram which shows at least **five** reasons why Rio de Janeiro is an important city.

You need to know:

- the social challenges facing Rio.

Student Book
See pages 154–5

What are the challenges?

Access to services

Health care

District	Zone	Infant mortality (per 1000)	Pregnant females getting medical care	Average life expectancy
Cidade de Deus	West	21	60%	45
Barra da Tijuca	West	6	100%	80
Rio de Janeiro		19	74%	63

Figure 1 *Health in two contrasting districts and Rio as a whole*

Education

Only half of all children continue their education beyond the age of 14. Reasons include:

- a shortage of schools and teachers
- a lack of money and a need for teenagers to work to support their families.

Access to resources

Water supply

- 12% of Rio's population had no running water.
- 37% is lost through leaks and illegal access.
- Droughts make water expensive.

Energy

- Frequent power cuts and blackouts.
- Many poorer people get their electricity by illegally tapping into the main supply.

How do solutions create opportunities?

Access to services

Health care

Medical staff detect and treat twenty different diseases in people's homes, reducing infant mortality and increasing life expectancy.

Education

The authorities have:

- given school grants to poor families
- opened a private university in Rocinha favela

Access to resources

Water supply

- 300 km new pipes and seven treatment plants built.
- By 2014, 95% of the population had mains water.

Energy

- 60 km of new power lines installed.
- a new nuclear generator built.

Six Second Summary

- Rio's urban growth creates challenges in providing health, education, clean water and energy.
- Improved access to services and resources creates social opportunities.

Over to you

Describe a) **two** challenges b) **one** opportunity for each of health care, education, water supply and energy in Rio.

You need to know:

- the economic opportunities and challenges facing Rio.

Student Book
See pages 156–7

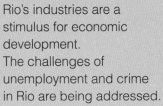

How have Rio's industries created economic development?

- Rio's industrial areas have boosted the city's economy.
- Rio provides more than 6% of Brazil's employment.
- Economic development has improved Rio's transport and environment.
- The city's favelas have improved.
- Large companies are now attracted to Rio.
- **Economic opportunities** have developed in the **formal economy**.

Big Idea

Rio's industries are a stimulus for economic development.
The challenges of unemployment and crime in Rio are being addressed.

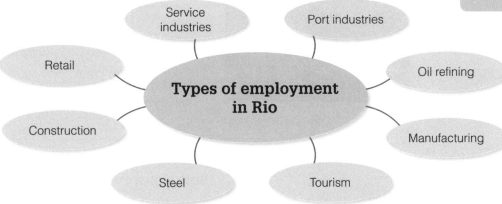

Types of employment in Rio: Service industries, Port industries, Oil refining, Manufacturing, Tourism, Steel, Construction, Retail

What are the challenges?

Unemployment

A recession in 2015 increased unemployment in Rio. There are wide contrasts in wealth.

Unemployment rates in favelas are over 20%. Most people work in the **informal economy** where jobs are poorly paid and irregular.

Crime

Murder, kidnapping and armed assault occur regularly. Powerful gangs control drug trafficking in many of the favelas.

What are the solutions?

Unemployment

- The Schools of Tomorrow programme aims to improve education in the poor and violent areas.
- Free child care is provided to enable teenage parents to return to education.

Crime

- In 2013, Pacifying Police Units (UPPs) were established to reclaim favelas from drug dealers.
- Police have taken control of some crime-dominated favelas.

Six Second Summary

- Rio's industrial areas have boosted the city's economy.
- The government is using education to try to reduce unemployment.
- The police are trying to take control away from criminals in the favelas.

Over to you

Either: Draw your understanding of the economic opportunities and challenges facing Rio. (No writing!)

Or: Create a mind map of the economic opportunities and challenges facing Rio.

Student Book
See pages 158–9

You need to know:

- how Rio is responding to its environmental challenges.

Air pollution and traffic congestion

Big Idea

Urban growth has created environmental issues which need to be managed.

Air pollution causes around 5000 deaths per year in Rio. Smog occurs in still conditions, when natural mist or fog mixes with vehicle exhaust fumes and pollutants from factories.

Traffic congestion increases stress and pollution. It happens because:

- steep mountains limit where roads can go
- the number of cars has grown
- high crime levels mean people prefer to drive.

Figure 1 *Traffic congestion in Rio*

Solutions

- Expansion of the metro system (cutting car use)
- New toll roads (so people think about the cost of travel)
- Making coast roads one-way during rush hours

Water pollution

- Guanabara Bay is highly polluted.
- Rivers are polluted by open sewers in the favelas because the government has not paid for sewage pipes.
- There have been oil spills from the Petrobras oil refinery.
- Ships empty their fuel tanks in the bay.

Figure 2 *Water pollution in Rio*

Solutions

- 12 new sewage works have been built since 2004.
- Ships are fined for discharging fuel illegally.
- 5 km of new sewage pipes have been installed.

Waste pollution

Many favelas are on steep slopes with few proper roads so waste collection is difficult. Most waste is dumped and pollutes the water system, causing diseases and encouraging rats.

Figure 3 *Waste in an exposed sewer canal, Tijuca district*

Solutions

A power plant has been set up which consumes 30 tonnes of rubbish a day and produces enough electricity for 1000 homes.

Six Second Summary

Ways of managing environmental issues include:
- reducing traffic congestion
- building new sewage works
- using rubbish in biogas power plants.

Over to you

Write down, with detail, **one** problem caused by, and **one** way of managing, each of these challenges:
- waste disposal
- water pollution
- air pollution
- traffic congestion.

Student Book
See pages 160–1

You need to know:

- about housing the poor in Rio.

Why have favelas grown?

In Brazil, **squatter settlements** or slums are called *favelas*. They are illegal settlements where people build homes on land they do not own, partly caused by high costs of housing in Rio.

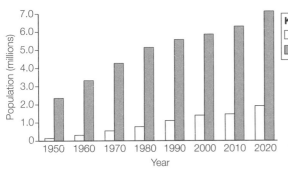

Key
- ☐ Favela population
- ▨ Total population

Figure 1 *The growth of the total and favela populations of Rio, 1950–2020*

What is Rocinha like?

Rocinha is the largest favela in Rio. It now has:

- 90% of houses built with brick, and with electricity, running water and sewerage systems
- bars, travel agents and shops (even McDonald's)
- schools, health facilities and university.

What are the challenges in squatter settlements?

Urban growth can create squatter settlements, where people face challenges.

Crime

- A high murder rate of 20 per 1000 people in many favelas.
- Drug gangs can dominate.

Health

- Infant mortality rates are as high as 50 per 1000.
- Waste cannot be disposed of and builds up in the street, increasing the danger of disease.

Services

In the non-improved favelas:

- 12% of homes have no running water
- 30% have no electricity
- 50% have no sewerage connections.

Construction

- Houses are built with basic materials on steep slopes.
- Heavy rain can cause landslides.

Unemployment

- Unemployment rates are as high as 20%.
- Average incomes may be less than £75 a month.

Figure 2 *A favela in Rio. Favelas may provide a better quality of life than the rural areas people came from.*

Six Second Summary

Squatter settlements cause challenges:
- poorly constructed houses
- lack of services and poor health
- high unemployment and crime

Over to you

Write a definition of a squatter settlement. Then list **seven** challenges that the people living there face. Consider which **two** challenges would be most difficult to overcome with **one** piece of supporting evidence.

Planning for Rio's urban poor

You need to know:

- how squatter settlements are being improved.

Student Book
See pages 162–3

Favela Barrio Project

This is a *site and service scheme* – the local authority provides land and services for residents to build homes.

Complexo do Alemão is a group of favelas in Rio's North Zone. The local authority has made many improvements.

✓ Paved roads
✓ Access to a water supply
✓ Improved **sanitation**
✓ A cable car system – inhabitants are given one free return ticket a day
✓ A Pacifying Police Unit (UPP), with police patrolling the community

How have the Olympics affected the favelas?

Some favelas were demolished to make way for the developments for the 2016 Olympic Games. The small town of Campo Grande saw 800 new houses being built.

😀 For some residents, the houses are better than the favelas.

☹️ Campo Grande lacks a sense of community, has no shops and is a 90-minute drive from the city centre.

Figure 2 Favelas were demolished to build new roads

Figure 1 Improvements in Complexo do Alemão

Has the Favela Barrio Project been a success or a failure?

😀 The quality of life, mobility and employment prospects of the inhabitants have improved.

There are still problems:

☹️ the newly built infrastructure is not being maintained

☹️ residents lack the skills and resources to make repairs

☹️ more training is needed to improve literacy and employment.

 Big Idea

This is an example of how urban planning is improving the quality of life for the urban poor.

 **Six Second Summary**

Improvements have been made to favelas:
- better roads
- better water supply
- a cable car system.

But, there are still problems.

 Over to you

Draw a table to show **four** strengths and **four** weaknesses in the schemes to improve housing.

Student Book
See pages 164–5

You need to know:

- the distribution of the population (how it is spread out) in the UK
- where the major cities in the UK are
- reasons why the population of the UK is distributed in this way.

How is the UK population distributed?

- The UK's population is unevenly distributed.
- 82% live in urban areas.
- A quarter of these urban dwellers live in London and the south-east of England.
- Many highland regions are very sparsely populated – upland areas are remote and can experience harsh weather conditions.

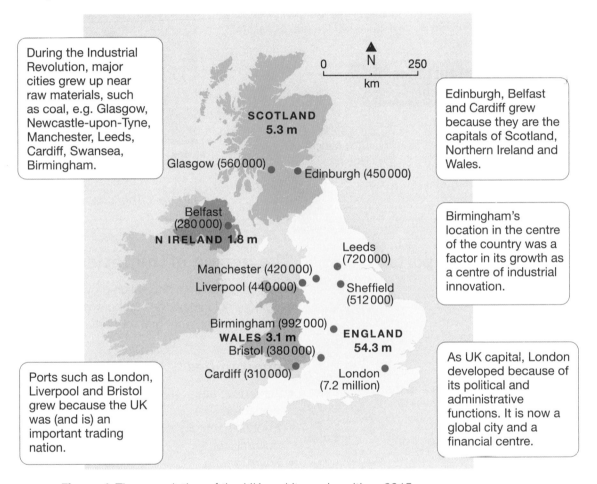

During the Industrial Revolution, major cities grew up near raw materials, such as coal, e.g. Glasgow, Newcastle-upon-Tyne, Manchester, Leeds, Cardiff, Swansea, Birmingham.

Edinburgh, Belfast and Cardiff grew because they are the capitals of Scotland, Northern Ireland and Wales.

Birmingham's location in the centre of the country was a factor in its growth as a centre of industrial innovation.

Ports such as London, Liverpool and Bristol grew because the UK was (and is) an important trading nation.

As UK capital, London developed because of its political and administrative functions. It is now a global city and a financial centre.

SCOTLAND 5.3 m
Glasgow (560 000)
Edinburgh (450 000)
Belfast (280 000)
N IRELAND 1.8 m
Leeds (720 000)
Manchester (420 000)
Liverpool (440 000)
Sheffield (512 000)
Birmingham (992 000)
WALES 3.1 m
ENGLAND 54.3 m
Bristol (380 000)
Cardiff (310 000)
London (7.2 million)

Figure 1 *The population of the UK and its major cities, 2015*

How might this distribution change?

- There has been a general drift towards London and south-east England.
- Immigrants generally settle in larger cities where there are more job opportunities.
- In recent years, there has been a movement from urban to rural areas. Many older people choose to retire near the coast or in the country.

Six Second Summary

- The UK's population is unevenly distributed.
- 82% live in urban areas.
- Cities grew near supplies of raw materials.
- There is now a general drift towards London and the south-east.

Over to you

On a blank map of the UK, mark the location of **nine** different cities. For at least **five** of those cities, write down why each one has grown.

Student Book
See pages 166–7

You need to know:

- where Bristol is located
- reasons why Bristol is important in both the UK and the wider world
- the impacts of national and international migration on a) the growth and b) the character of Bristol.

What makes Bristol a major UK city?

Bristol is the largest city in south-west England. It is important regionally and nationally.

 **Big Idea**

Bristol is a case study of a major city in the UK.

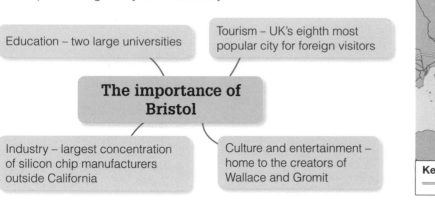

Education – two large universities

Tourism – UK's eighth most popular city for foreign visitors

The importance of Bristol

Industry – largest concentration of silicon chip manufacturers outside California

Culture and entertainment – home to the creators of Wallace and Gromit

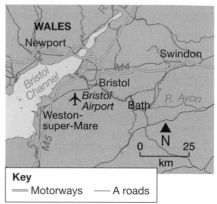

Key
— Motorways — A roads

Figure 1 *The location of Bristol*

Why is Bristol an important international city?

Transport

- Good road and rail links
- Ferry services to Europe
- Two major docks
- Bristol airport links to Europe and the USA

Industry

- Global industries like aerospace and media
- Inward investment from abroad

Education

- Attracts international students

The impact of migration

About half of Bristol's population growth comes from migration, including large numbers from EU countries. Migration has brought both opportunities and challenges.

Opportunities

- A hard-working workforce
- Enriching the city's cultural life
- Mainly young migrants help to balance the ageing population

Challenges

- Housing provision has not kept pace with population growth – so Bristol is very expensive for housing rental or purchase
- Teaching children whose first language is not English
- Integration into the wider community

 Six Second Summary

- Bristol is important in the UK, e.g. education and tourism.
- Bristol is important in the wider world, e.g. industry and transport.
- Migration has brought both opportunities and challenges.

 Over to you

Under the headings **education**, **industry** and **culture**, explain:

- why Bristol is an important city
- the impact of migration on Bristol.

*Student Book
See pages 168–9*

You need to know:

- what urban changes are affecting Bristol
- how those urban changes have created social (cultural) opportunities.

Changes in Bristol

- Bristol's population is increasing.
- Its population is becoming more ethnically diverse.
- It has good transport links – good for business and commuters.
- Over 2 million people live within 50 km of the city.

Social (cultural) opportunities

Entertainment

There are nightclubs, bars and a vibrant underground music scene. Colston Hall is a venue for concerts and entertainment. Theatres include the Bristol Old Vic. Migrants contribute to music, art, literature and food.

Figure 1 *Bristol's Harbourside is a vibrant mix of social and cultural venues*

Sport

Sports teams are developing their stadia to provide a range of leisure and conference facilities. This often involves new stadia on the outskirts of the city.

Shopping

Bristol has seen major changes. The retail park at Cribbs Causeway affected the outdated Broadmead shopping development in the city centre. As a result, Cabot Circus was developed.

Figure 2 *The interior of Cabot Circus Shopping Centre*

Cabot Circus

- Opened in 2008 at a cost of £500 million.
- Shops and leisure facilities take up two-thirds of its floor space.
- There are also offices, a cinema, a hotel and 250 apartments.

Bristol's Harbourside

- Conversion of workshops and warehouses into bars, nightclubs and cultural venues.
- Includes an art gallery, museum and the At-Bristol science centre.
- The Harbourside Festival attracts around 300 000 spectators.

Six Second Summary

Urban change in Bristol has created:
- social and cultural opportunities
- improved sports facilities
- improved shopping facilities.

Over to you

List **three** changes that have affected Bristol. Draw lines to link each one with a different social opportunity that has occurred.

Student Book
See pages 170–1

You need to know:

- how Bristol's industry has changed (this is an urban change)
- how that change has created economic opportunities, such as employment.

How has Bristol's industry changed?

The closure of Bristol's port meant its industry changed. Since then, major developments have been in **tertiary (services)** and quaternary (high-tech) sectors. This creates employment.

High-tech businesses have been attracted to Bristol because of:

- a government grant of £100 million to become a Super-Connected City with high-speed broadband
- a university-educated and skilled workforce
- advanced research at the university and in local IT and aerospace industries

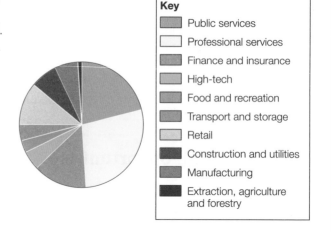

Key
- Public services
- Professional services
- Finance and insurance
- High-tech
- Food and recreation
- Transport and storage
- Retail
- Construction and utilities
- Manufacturing
- Extraction, agriculture and forestry

Figure 1 *Bristol's employment structure*

Industries in Bristol include:

Defence Procurement Agency (DPA)

- Employs over 10000 people.
- Supplies the army, air force and navy with everything from boots to aircraft carriers.
- Established on a **greenfield site**.
- Created a demand for housing which contributed to the city's **urban sprawl**.

The aerospace industry

- Fourteen of the fifteen main global aircraft companies are found in the Bristol region, including Rolls-Royce, Airbus and GKN Aerospace.
- Supply chains have grown up in the region to supply these high-tech companies.

Aardman Animations

- Became well-known for using stop-motion clay animation techniques.
- Entered the computer animation market.
- Won an Oscar and many other awards.

Figure 2 *The Aardman characters Wallace and Gromit*

Figure 3 *The purpose-built DPA headquarters at Filton*

Six Second Summary

- A growing number of people in Bristol are employed by high-tech companies.
- Employment is an economic opportunity created by the growth of high-tech industries.

Over to you

Write your top **five** opportunities that urban change in Bristol has created. They can be social or economic.

Student Book
See pages 172–3

You need to know:

- what urban changes are affecting Bristol's environment
- how those changes have created further opportunities.

How are changes affecting Bristol's environment?

- In 2015, Bristol became the first UK city to be awarded the status of European Green Capital.
- It also plans to increase the number of jobs in low-carbon industries.

Increase the use of renewable energy from 2% (2012)

Improve energy efficiency – reduce energy use by 30% and CO_2 emissions by 40% by 2020

Reduce water pollution by improved monitoring and maintenance

BRISTOL 2015 EUROPEAN GREEN CAPITAL

Establish an Air Quality Management plan to monitor air pollution

Increase the use of **brownfield sites** for new businesses and housing

Figure 1 *What is Bristol doing to improve the environment?*

Big Idea

Urban greening is an example of an environmental opportunity. *Integrated transport systems* can also provide social and economic opportunities.

Add a WOW! factor

Make sure you know which opportunities are economic, social and environmental. You may get asked a question which includes more than one category.

How has this urban change created opportunities?

An integrated transport system (ITS) for Bristol

An ITS connects different methods of transport. It encourages people to switch to using public transport.

- The Rapid Transit Network – three bus routes linking the railway station to Park and Ride sites.
- Electrification of the railway line to London – greener and more reliable journeys.
- Aims to double the number of cyclists by 2020.

Urban greening

Urban greening is the process of increasing and preserving open space in urban areas.

- More than a third of Bristol is open space.
- There are eight nature reserves and 300 parks in the city.
- Queen Square was once a dual carriageway, but is now an open space with cycle routes.

Figure 2 *Queen Square, Bristol*

Six Second Summary

- Bristol's economy has focused on green issues.
- Integrated transport systems provide social, economic and environmental opportunities.
- Urban greening is another environmental opportunity.

Over to you

Produce a Venn diagram of social, economic and environmental opportunities in Bristol. Explain why there may be cross-over between categories for some of the opportunities.

Student Book See pages 174–5

You need to know:

- how changes in Bristol have created environmental challenges.

 Big Idea

Changes in Bristol
- Movement of the port downstream from the city
- Bristol's population is growing rapidly

have created →

Environmental Challenges
- Many industrial buildings are now **derelict**.
- Demand for new homes has led to *urban sprawl* on the **rural-urban fringe** and building on *brownfield* and *greenfield sites*.

Dereliction: a challenge

Problems in Stokes Croft

- An inner-city area with housing once built for industrial workers.
- Housing became derelict.
- There were problems with squatters, riots and antisocial behaviour.

Figure 1 *Graffiti art on abandoned building in Stokes Croft*

What is being done to improve the area?

- Bristol City Council obtained lottery grants to help improve the area.
- Activists and artists want to revitalise the area through community action and public art, including graffiti art.

Urban Sprawl: a challenge

How has urban growth led to urban sprawl?

Bristol needs new housing because of:

- a rapidly growing population
- demolition of older slum dwellings.

Urban sprawl has extended, particularly to the north-west of the city. The new town of Bradley Stoke has extended the city to the north.

Figure 2 *Bristol's Harbourside – a brownfield development*

What is being done to reduce urban sprawl?

- Bristol is developing brownfield sites such as Harbourside (see 14.3).
- Between 2006 and 2013, only 6% of new housing developments were on greenfield land.

 Six Second Summary

- Environmental challenges in Bristol include dereliction and building on brownfield and greenfield sites.
- Urban sprawl has an impact on the rural–urban fringe.
- Building on brownfield sites can reduce urban sprawl.

 Over to you

Write **three** questions about the material on this page so that you could test a friend. You have to know the answers too!

Creating a clean environment in Bristol

Student Book
See pages 176–7

You need to know:

- that a growing population in Bristol is an example of an urban change
- how that growing population has created problems of waste disposal and atmospheric pollution.

Waste disposal

The amount of waste produced per head in Bristol is 23% lower than the UK average, but:

- Bristol produces half a million tonnes of waste per year
- it is among the worst cities in the country in terms of the amount of food waste it produces.

How is Bristol reducing the environmental impact of waste disposal?

- Income is generated when recycled materials are sent to reprocessing plants.
- The Avonmouth waste treatment plant treats 200 000 tonnes of waste per year.
- Any non-recyclable waste is used to generate electricity. It supplies nearly 25 000 homes in the Bristol area.

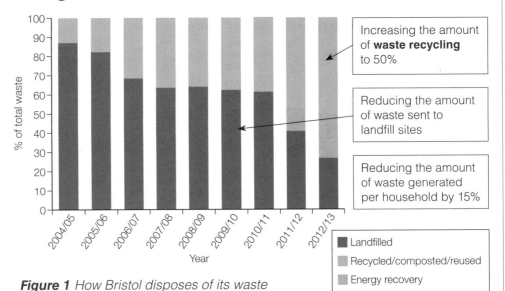

Increasing the amount of **waste recycling** to 50%

Reducing the amount of waste sent to landfill sites

Reducing the amount of waste generated per household by 15%

Figure 1 *How Bristol disposes of its waste*

Atmospheric pollution

- Vehicle emissions are the main cause.
- An estimated 200 people die prematurely each year in Bristol, as a result of air pollution.

Actions to improve the air quality include:

- the Frome Gateway, a walking and cycling route to the city centre
- an electric vehicle programme
- a smartphone app with information about public transport services, connections and delays.

Add a *WOW!* factor

Look out for trends and patterns when describing a graph. Use words like 'increases', 'decreases' or 'fluctuates'.

Six Second Summary

Environmental challenges in Bristol are:
- waste disposal
- atmospheric pollution.

Over to you

Cover up everything on this page apart from the divided bar graph. Use it to describe how Bristol disposes of its waste.

**Student Book
See pages 178–9**

> **You need to know:**
>
> - how urban change has created social and economic challenges – urban deprivation and inequalities in housing, education, health and employment.

Inequality in Bristol

Lack of investment in Bristol has led to **inequalities** between some areas such as Filwood, which has high levels of **social deprivation**, and Stoke Bishop, which is a more affluent area.

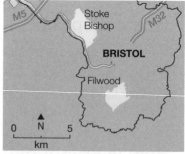

Figure 1 *The location of Filwood and Stoke Bishop in Bristol*

Filwood: an area containing urban deprivation

Housing

- Homes are owner-occupied or rented from the city council.
- Many homes are poorly insulated.

Education

- In 2013, only 36% of students got top grades at GCSE.

Health

- Life expectancy is 78 years, lower than the UK average.
- Death rates from cancer are above the UK average.
- It has Bristol's lowest participation rates in active sport and creative activities.

Employment and economic

- One-third of people aged 16–24 are unemployed.
- Over half of all children live in low-income households

Figure 2 *Empty shops in Filwood*

Stoke Bishop: a very affluent suburb

Housing

- 81% of the housing is owner-occupied.
- Includes Sneyd Park, which is home to many millionaires who live in large Victorian and Edwardian villas.

Education

- Nearly 50% of the population have a degree or equivalent.
- 94% of 16-year-olds got the highest grades in five or more GCSEs.

Health

- Life expectancy is 83 years, above the UK average.
- Death rates are better than the Bristol and UK averages.

Employment and economic

- Only 3% of people are unemployed.
- Fewer than 4% of children live in poverty.

Figure 3 *The centre of Stoke Bishop*

 Six Second Summary

- Filwood is an area of Bristol containing urban deprivation.
- In contrast, Stoke Bishop is an affluent area.
- There are housing, education, health and economic inequalities between those two areas.

Over to you

Give each of four cards one of the labels 'Housing', 'Education', 'Health' or 'Economic'. On each one, write contrasting facts about Filwood and Stoke Bishop.

New housing for Bristol

Student Book
See pages 180–1

You need to know:

- that urban change has created two environmental challenges – building on a) greenfield and b) brownfield sites
- neighbouring towns have become commuter settlements, creating traffic congestion.

Urban sprawl, the rural–urban fringe and commuter settlements

Nearby towns, such as Wotton-under-Edge and Clevedon, have expanded to become *commuter settlements*.

The green belt was set up to prevent urban sprawl on the *rural–urban fringe*.

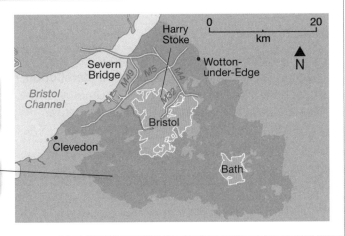

Figure 1 *The green belt around Bristol*

Housing developments on greenfield land

Harry Stoke

A new development of 1200 homes has been built on land at Harry Stoke, with 2000 more planned to be completed by the end of 2017.

Local people objected. They were concerned about:

- traffic congestion, noise and poor air quality
- loss of animal habitats
- the effect on the local flood risk.

Housing development on brownfield land

Between 2006 and 2013, 94% of new housing was built on brownfield sites.
There is a growing need for brownfield land for student accommodation.

Bristol Harbourside

When Bristol's port closed, so did several industries around the docks. Empty industrial buildings were regenerated for housing and cultural facilities.

Advantages of the scheme

- A very run-down area of the city has been redeveloped.
- People still live in the centre.

Disadvantages of the scheme

- Not everyone is happy about the architecture.
- Renovation has been costly, so flats are expensive.

Figure 2 *New apartments on Bristol Harbourside*

Six Second Summary

- Commuter settlements have grown.
- Local people sometimes object to greenfield schemes.
- There are advantages and disadvantages to building on brownfield sites.

Over to you

Decide whether it would be better to build on greenfield or brownfield sites. Explain your decision to a friend, making sure you include arguments about each type of site.

The Temple Quarter regeneration (1)

Student Book
See pages 182–3

Example

You need to know:

- the reasons why the Temple Quarter needed regeneration
- what the main features of the project are.

Developing brownfield sites

Advantages

- Existing buildings can be put to a range of uses.
- The land is often disused or in a state of **dereliction**, so any changes to the land are usually an improvement.
- Sites are usually in urban areas, so urban sprawl and also car use for commuting are reduced.

Disadvantages

- Expensive to build on – old buildings may need demolishing.
- Sites may be contaminated from previous industrial use.

Why did the Temple Quarter need regeneration?

- The area was very run down.
- It gave a bad impression to visitors driving in from the south or south-east, or arriving at Temple Meads railway station.

There are four separate areas within the Temple Quarter.

Big Idea

The Temple Quarter is an example of an **urban regeneration** project.

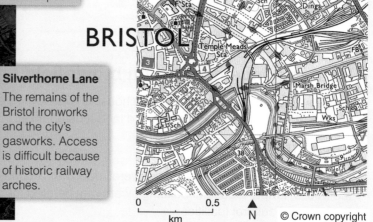

Avon Riverside

Old industrial buildings used for light industries. Green spaces were created by demolishing a former diesel depot.

Temple Meads City Gateway

Dominated by Temple Meads railway station. Cut off from the rest of the area by the Temple Gate dual carriageway built in the 1970s.

Silverthorne Lane

The remains of the Bristol ironworks and the city's gasworks. Access is difficult because of historic railway arches.

Temple Quay

A former industrial area, including ropeworks, timber yards and potteries.

Figure 1 *Aerial view of the Temple Quarter in the 1990s*

BRISTOL

Figure 2 *OS map of the Temple Quarter area of Bristol (shows the same area as Figure 1)*

Six Second Summary

The Temple Quarter was an area that needed regeneration because:

- it was very run down
- it was a former industrial area
- it gave visitors to the city a bad initial first impression.

Over to you

Cover everything on this page apart from the OS map. Point out features on the map to show why the Temple Quarter needed regeneration.

Student Book
See pages 184–5

You need to know:

- what the main features of the urban regeneration project in the Temple Quarter are.

Example

How has the area been regenerated?

The target is to create 17 000 new jobs by 2037, focusing on several key projects (Figure **2**).

Improved access from in and around Bristol
- Improvements to Temple Meads station.
- Improved road layout with links to the rapid transit network and the Bristol–Bath cycle path.

Enterprise Zone status
Enterprise Zones offer incentives to businesses to move to the area, including low rents and business taxes.

Temple Quarter Regeneration

New bridge across the River Avon
Gives access to the new Bristol Arena, currently being built.

Figure 1 *Key aspects of the Temple Quarter regeneration project*

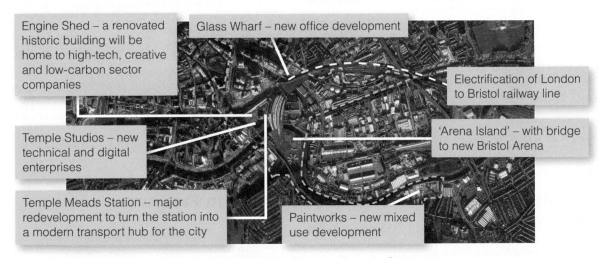

Engine Shed – a renovated historic building will be home to high-tech, creative and low-carbon sector companies

Glass Wharf – new office development

Electrification of London to Bristol railway line

'Arena Island' – with bridge to new Bristol Arena

Temple Studios – new technical and digital enterprises

Temple Meads Station – major redevelopment to turn the station into a modern transport hub for the city

Paintworks – new mixed use development

Figure 2 *Key regeneration projects in Bristol's Temple Quarter*

Bristol Arena (under construction)

Access will be by the new bridge over the river (see above), as well as by a pedestrian and cycle bridge. This route is to be redeveloped with cafés, offices and flats.

The arena will be used for concerts, conventions, exhibitions and sporting events. The area around the arena will host outdoor events.

Figure 3 *An artist's impression of Bristol Arena*

 Six Second Summary

The main features of the Temple Quarter regeneration include:
- improved transport links
- encouraging more visitors, new businesses and employment
- the Bristol Arena.

 Over to you

List **two** ways in which urban change in Bristol **has**, and two ways in which it **has not** created social and economic challenges.

Student Book
See pages 186–7

> **You need to know:**
>
> • that sustainable development of urban areas requires social, economic and environmental planning.

Urban sustainability

In cities, huge quantities of energy and water are consumed. Waste disposal and traffic congestion are also problems.

Cities can tackle these problems and become more **sustainable**.

Freiburg in Germany is an example of a sustainable city.

 Big Idea

Sustainability refers to actions that meet the needs of the present without reducing the ability of future generations to meet their own needs.

Environmental planning in Freiburg

Waste recycling is a key feature of sustainable urban living.

...reducing annual waste disposal from 140 000 to 50 000 tonnes in 12 years

...recycling more than 88% of packing waste

Freiburg has reduced landfill by...

...provided energy for 28 000 homes from burning non-recyclable waste

...building a biogas digester for organic food

Environmental planning also involves the use of brownfield sites.

Social planning in Freiburg

In Freiburg, local people are involved in urban planning at both local and city level.

• Local people can invest in renewable energy resources.
• Financial rewards are given to people who compost their green waste and use textile nappies.

Economic planning in Freiburg

Freiburg is a city where people come to attend conferences on sustainability, and this provides jobs for local people.

More than 10 000 people are employed in 1500 environmental businesses in the city.

Figure 1 *Freiburg's Solar Factory*

A Solar Training Centre provides training in the skills needed for the new solar technology

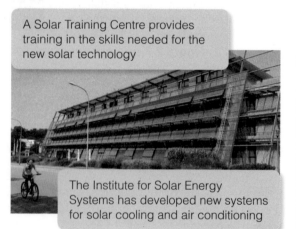

The Institute for Solar Energy Systems has developed new systems for solar cooling and air conditioning

 Six Second Summary

• Freiburg is an example of a sustainable city.
• Waste recycling is a feature of sustainable urban living.
• Sustainability requires social, economic and environmental planning.

 Over to you

Draw a spider diagram to show the strategies used to make Freiburg sustainable. On each 'leg' of the spider, explain why each strategy makes Freiburg more sustainable.

*Student Book
See pages 188–9*

You need to know:

- how urban water supply, energy and green spaces can be made sustainable.

Three features of sustainable living

1

Water conservation

Freiburg's waste water system allows rainwater to be retained, reused or to seep back into the ground. Water conservation involves:

- collecting rainwater for use indoors
- green roofs
- unpaved tramways
- pervious pavements that allow rainwater to soak through.

Figure 1 Green roofs look attractive and are used to harvest rainwater

2

Energy conservation

Freiburg has a strict energy policy based on:

- energy saving
- efficient technology
- use of *renewable energy sources*.

Freiburg is one of the sunniest cities in Germany so solar power is an important form of renewable energy. There are about 400 solar panel installations in the city.

Figure 2 Freiburg's Solar Settlement and Solar Business Park

The largest proportion of Freiburg's renewable electricity comes from **biomass** using waste wood and rapeseed oil. Biogas is produced from organic waste. This produces enough energy to heat Freiburg's three swimming pools!

3

Creating green space

Green spaces:

- help keep the air clean
- provide a natural and free recreational resource.
- provide a habitat for wildlife.

Figure 3 Freiburg – the 'green city'

40% of the city is forested

In the Riselfeld District only 78 hectares are built on, leaving 240 hectares of open space

The River Dreisam is not managed – it provides natural habitats for flora and fauna

44 000 trees have been planted in parks and streets

Only native trees and shrubs are planted in the 600 hectares of parks

Six Second Summary

Features of sustainable living include:
- water conservation
- energy conservation
- creating green space.

Over to you

- Write out the three bullet points from the Six Second Summary as headings. Under each one write **three** sentences to explain how that feature can lead to sustainable urban living. Add pictures/sketches to aid memory.

Student Book
See pages 190–1

You need to know:

- how urban transport strategies can reduce traffic congestion.

Reducing traffic congestion

Bristol's plans to tackle traffic congestion are explained in 14.5. Here are three more examples:

1

Freiburg

Freiburg has an integrated traffic plan.

The tram network provides efficient, cheap and accessible public transport.

There are also:

- 400 km of cycle paths
- restrictions on car parking spaces; in Vauban district each one costs £20 000!

Low fares allow unlimited travel

The tram network is connected to the bus routes

70% of the population live within 500 m of a tram stop

Figure 1 Advantages of Freiburg's tram network

2

Singapore

Restricted entry to the city centre during rush hours

Electronic road pricing on major roads

Measures to reduce the volume of traffic

High petrol prices

An overhead railway system and efficient bus network

As a result of these measures:

- there is 45% less traffic and 25% fewer accidents in the city centre
- two-thirds of all daily journeys are now by public transport.

3

Beijing

The restrictions in Beijing have led to a 20% drop in car use. But building roads has also increased car use at the expense of cycling.

Figure 2 Traffic management strategies in Beijing

Cars are banned from the city one day a week, based on a number plate system

Congestion charge and pollution tax introduced

Only 20% of people who apply to own a vehicle are allowed to do so

Thirty new metro lines and a rapid bus transit system to be built by 2020

 Six Second Summary

Sustainable traffic management strategies include:
- improving public transport
- restricting use of vehicles
- adding charges and taxes to make driving expensive.

 Over to you

List the bullet points from the Six Second Summary. Underneath each point, write **two** examples of that particular strategy.

Section B
The changing economic world

Your exam

Section B The changing economic world makes up part of Paper 2: Challenges in the human environment.

Paper 2 is a one-and-a-half hour written exam and makes up 35 per cent of your GCSE. The whole paper carries 88 marks (including 3 marks for SPaG) – questions on Section B will carry 30 marks.

You need to study all the topics in Section B – in your final exam you will have to answer questions on all of them.

Tick these boxes to build a record of your revision

Your revision checklist

Spec key idea	Theme	1	2	3
16 The development gap				
There are global variations in economic development and quality of life	16.1 Our unequal world			
	16.2 Measuring development			
	16.3 The Demographic Transition Model			
	16.4 Changing population structures			
	16.5 Causes of uneven development			
	16.6 Uneven development – wealth and health			
	16.7 Uneven development – migration			
Various strategies exist for reducing the global development gap	16.8 Reducing the gap			
	16.9 Reducing the gap – aid and intermediate technology			
	16.10 Reducing the gap – fair trade			
	16.11 Reducing the gap – debt relief			
	16.12 Reducing the gap – tourism			
17 Nigeria: a newly emerging economy				
Some LICs or NEEs are experiencing rapid economic development which leads to significant social, environmental and cultural change	17.1 Exploring Nigeria (1)			
	17.2 Exploring Nigeria (2)			
	17.3 Nigeria in the wider world			
	17.4 Balancing a changing industrial structure			
	17.5 The impacts of transnational corporations			
	17.6 The impacts of international aid			
	17.7 Managing environmental issues			
	17.8 Quality of life in Nigeria			
18 The changing UK economy				
Major changes in the economy of the UK have affected, and will continue to affect, employment patterns and regional growth	18.1 Changes in the UK economy			
	18.2 A post-industrial economy			
	18.3 UK science and business parks			
	18.4 Environmental impacts of industry			
	18.5 Changing rural landscapes in the UK			
	18.6 Changing transport infrastructure (1)			
	18.7 Changing transport infrastructure (2)			
	18.8 The north-south divide			
	18.9 The UK in the wider world (1)			
	18.10 The UK in the wider world (2)			

Student Book
See pages 194–5

- there are global variations in economic development and quality of life.

What is development?

Development means positive change that makes things better. It usually means that people's standard of living and quality of life will improve.

The **development gap** is the difference in standard of living between the world's richest and poorest countries.

Remember!

These statistics give broad measures for countries rather than individuals. Quality of life also considers, for example, safety and security, freedom and the right to vote.

Measuring development

Gross National Income (GNI)

- **GNI** is an economic measure of development.
- It is the total value of goods and services produced by a country, plus money earned from, and paid to, other countries.
- It is expressed as per head (per capita) of the population.

Some countries (NEEs) have begun to experience higher rates of economic development. For example, the BRICS (Brazil, Russia, India, China, South Africa) and MINT (Mexico, Indonesia, Nigeria, Turkey) countries.

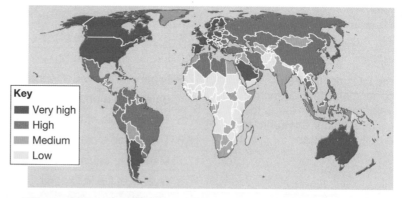

Key
- High
- Higher middle
- Lower middle
- Low
- No data

Figure 1 *Gross National Income per capita in PPP terms, 2013. PPP means Purchasing Power Parity – it shows GNI in terms of what it will buy using local prices.*

Human Development Index (HDI)

HDI is a social measure that is expressed in values 0–1, where 1 is the highest. It considers:

- **life expectancy** at birth
- number of years of education
- GNI per head.

Key
- Very high
- High
- Medium
- Low

Figure 2 *World HDI scores, 2014*

- Development is a change that generally makes people's lives better.
- GNI per head is an economic measure of development.
- HDI is a social measure of development.

- Write clear definitions of development, Gross National Income (GNI) and Human Development Index (HDI).
- Name the members of the a) BRICS b) MINT countries.
- Explain the difference between GNI and HDI.

You need to know:

- how useful economic and social measures (or indicators) of development are
- examples of these measures
- the limitation of these measures.

*Student Book
See pages 196–7*

How useful are measures of development?

Birth rate

As a country develops, women become more educated and want a career. They marry later and have fewer children.

Death rate

Developed countries tend to have older populations resulting in a high **death rate**. Less developed countries may have very low death rates because proportionally more young people have survived their early years.

Infant mortality

A useful measure of a country's health care system.

Literacy rate

A high **literacy rate** means a good education system.

	HICs	LICs
Birth rate	Low	High
Death rate	Depends on the ages of the population and health care availability	
Infant mortality rate	Low	High
Literacy rate	High	Low

Figure 1 How measures of development vary with economic development

Country	GNI per head (US$)	HDI	Birth rate (per 1000 per year)	Death rate (per 1000 per year)	Infant mortality (per 1000 live births per year)	Literacy rate (%)
UK	43430	0.907	12.17	9.35	4.38	99.0
China	7400	0.727	12.49	7.53	12.44	96.4
Nigeria	**2970**	**0.514**	**37.64**	**12.90**	**72.70**	**59.6**
Bangladesh	1080	0.570	21.14	5.61	44.09	61.5
Zimbabwe	840	0.509	32.26	10.13	26.11	86.5

Figure 2 Measures of development for selected countries

Limitations of economic and social measures

A single measure of development can give a false picture, as it gives the *average* for the whole country.

Other factors limit the usefulness of development measures:

- Data could be out of date, unreliable or hard to collect.
- Data may not take into account subsistence or informal economies.

Add a WOW! factor

Make sure you understand what each measure means and how it shows a country's level of development. You are not expected to memorise all the data but two or three examples could be useful. Trends in the data are more important.

Six Second Summary

- Economic and social measures can show how developed a country is.
- There may be limitations to the usefulness of data if they are out of date or unreliable.

Over to you

Cover up everything on this page except for the table of data. Explain to somebody in your home what each measure shows about the level of development in those countries.

The Demographic Transition Model

Student Book
See pages 198–9

- how levels of development can be linked to the Demographic Transition Model (DTM).

The Demographic Transition Model (DTM) shows changes over time in the population of a country.

The total population responds to variations in birth and death rates (natural change). It is also affected by migration. Migration is not shown on the DTM.

As a country becomes more developed, its population characteristics change.

Stage 1

- High birth rate
- High death rate
- Both fluctuate because of disease, famine and war
- Population fairly stable

Example: Traditional rainforest tribes with little contact with the outside world. There are now no Stage 1 countries in the world.

Stage 2

- Death rate decreases
- Birth rate remains high
- Population grows

Example: Afghanistan – many poor countries are in Stage 2.

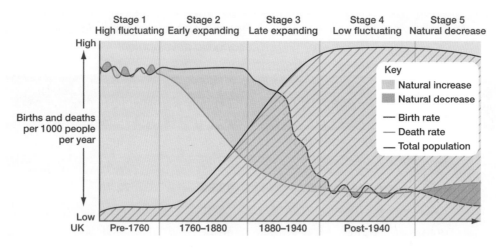

Figure 1 What links the DTM with development?

Stage 3

- Birth rate drops rapidly
- Death rate continues to decrease but more slowly
- Population still grows, but not quite as fast

Example: Nigeria – an NEE experiencing economic growth.

Stage 4

- Low birth rate
- Low death rate
- Birth rate can fluctuate depending on the economic situation

Example: USA – one of the most developed countries in the world, with good health care and women who pursue careers.

Stage 5

- Birth rate falls below death rate
- Death rate increases slightly because of ageing population
- Population decreases unless immigration replaces the retired population

Examples: Japan and Germany – well-developed countries with an ageing population.

Six Second Summary

- The DTM shows changes in birth rate, death rate and total population.
- As a country becomes more developed, these characteristics change.

Over to you

Draw the **five** stages of the DTM from memory and annotate it to show your understanding.

*Student Book
See pages 200–1*

You need to know:

• how the population structures of two contrasting countries are changing.

Population pyramids and the DTM

Countries at different stages of the DTM have population pyramids of different shapes.

Taller top indicates increasing life expectancy

Narrowing base indicates reducing birth rates

Wider middle indicates lowering death rates

Figure 1 *Population pyramids for the stages of the DTM*

Mexico's changing population structure (Stage 3)

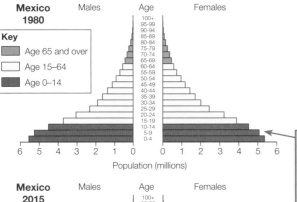

Key
- Age 65 and over
- Age 15–64
- Age 0–14

A wide base shows a large proportion of young people.

The bars are wider than 1980 showing the death rate is falling.

No more 'steps' at the bottom of the base show the birth rate is falling.

Japan's changing population structure (Stage 5)

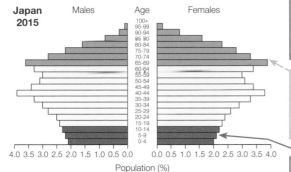

Wide bars at the top show people are living longer.

A narrowing base shows a falling birth rate.

Japan's total population is getting smaller.

The dependency ratio

This is the proportion of people below (aged 0–14) and above (over 65) normal working age. The lower the number, the greater the number of people who work.

Big Idea

Population structure considers how the number of men and women in different age groups is changing. It is studied using *population pyramids*.

Six Second Summary

• Population pyramids show population structure by age and gender.
• Pyramids change with different stages of the DTM.

Over to you

Using your drawing of the DTM from 16.3, add diagrams of population pyramids appropriate for each stage.

Student Book
See pages 202–3

You need to know:

- the physical, economic and historical causes of uneven development.

Physical

- Landlocked countries are cut off from seaborne trade, which is important for economic growth.
- Climate-related diseases and pests affect the ability of the population to stay healthy enough to work.
- Extreme weather, such as cyclones, droughts and floods, can slow development and it can be costly to repair damaged infrastructure.
- Lack of adequate supplies of clean water can affect farming and the health of workers.

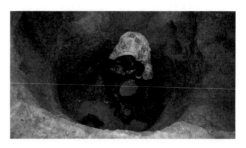

Figure 1 *Clean water is important for development*

Economic – trade

- Rich countries and large international companies want to pay as little as possible for their raw materials – many of which come from LICs.
- Supply of raw materials often outstrips demand, which keeps prices low.
- Processing (which adds value) takes place in richer countries.
- The rich countries get richer and the poorer countries are less able to develop.
- LICs and NEEs have traditionally exported *primary products*; although in the last 20 years, some have developed manufacturing. Manufactured products now make up about 80% of NEE exports.
- The price of raw materials fluctuates a lot (e.g. copper, see Figure **2**). In Zambia, copper accounts for over 60% of the total value of exports.

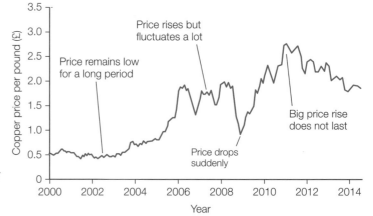

Figure 2 *World price for copper 2000–14*

Historical – colonialism

- Almost all the wealth produced during the colonial period (around 1650–1950) went to European powers.
- Since 1950 former European colonies have gained independence.
- Independence has often been a difficult process, resulting in civil wars and political struggles for power, which has continued to hold back development.

Six Second Summary

Causes of uneven development can be:

- physical (e.g. extreme weather or landlocked countries)
- economic (e.g. trade conditions)
- historical (e.g. colonialism).

Over to you

Talk for one minute about the causes of uneven development. Name and explain each cause, and give examples.

Student Book
See pages 204–5

You need to know:

- how uneven development leads to inequalities of wealth and health.

Imbalances between rich and poor

Imbalances exist between countries. Some countries have lower levels of development and a poorer quality of life than others.

Imbalances also exist *within* countries. Areas of considerable poverty can be found in rich countries, and great wealth in areas of poor countries.

Figure 1 *A cartoon highlighting global inequalities*

Disparities in wealth

- In 2014, the fastest growth of wealth was in North America, which now holds 35% of total global wealth.
- Of the NEEs, China has recorded the highest growth since 2000.
- Africa's share of global wealth remains very small (about 1%).

Figure 2 *Population and wealth by region, 2014*

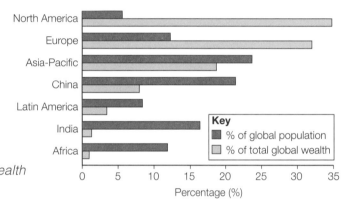

Key
- ■ % of global population
- □ % of total global wealth

Disparities in health

Levels of development are closely linked to health. In poor countries, health care is often patchy.

Children under 15 years account for 4 in every 10 deaths; the over 70s account for only 2 in 10 deaths.

Complications of childbirth are one of the main causes of death amongst under 5 year olds.

The over 70s account for 7 in every 10 deaths.

Main causes of death are chronic diseases, e.g. heart disease or cancer.

Low-income countries

High-income countries

Infectious diseases are the main cause of death, e.g. lung infections, HIV/AIDS and malaria.

Lung infections are the only main infectious cause of death.

Only 1 in every 100 deaths is among children under 15 years.

 Six Second Summary

- Uneven development leads to disparities in wealth and health.
- Lower levels of development can affect the causes of death in different countries.

Over to you

- Create a mnemonic to help you learn the reasons for disparities in health.
- In two days' time, see how much you have remembered about this page before looking at it again.

You need to know:

- how uneven development leads to international migration.

Student Book see pages 206–7

What are the different types of migration?

Migration is the movement of people from place to place. It can be voluntary or forced. International migration is a consequence of uneven development, as people seek to improve their quality of life.

The following terms are important.

Immigrant – a person who moves into a country.

Emigrant – a person who moves out of a country.

Economic migrant – a person who moves voluntarily to seek a better life, such as a better-paid job.

Refugee – a person forced to move from their country of origin, often as a result of civil war or a natural disaster.

Displaced person – a person forced to move from their home but who stays in their country of origin.

Middle East refugee crisis, 2015

In Syria a civil war has raged since 2011 causing four million people to flee the country to temporary camps in Turkey, Jordan and Lebanon.

Thousands have made the dangerous journey across the Mediterranean with the loss of many lives. Some people travelled by land through Turkey and into Eastern Europe. An estimated 1.1 million migrants entered Germany in 2015.

In March 2016, the **European Union (EU)** and Turkey signed a deal to give Turkey political and financial benefits in return for taking back migrants.

Figure 1 *Main migration routes from Syria into Europe*

Economic migration into the UK

- Since 2004 over 1.5 million economic migrants have moved to the UK, two-thirds of whom are Polish.
- Most migrants pay tax and work hard.
- Migrants can put pressure on services such as health and education.

Figure 2 *Migrant workers on a Lincolnshire farm*

Six Second Summary

- A consequence of uneven development is international migration.
- There are different types of migration.

Over to you

List as many causes and consequences of uneven development as you can.

Reducing the gap

You need to know:

- how investment, industrial development and tourism can reduce the development gap.

*Student Book
See pages 208–9*

What strategies can reduce the development gap?

Investment

Many countries and TNCs invest money and expertise in LICs (to increase their profits), which supports the LICs development by providing employment and income.

- Chinese companies have invested billions of dollars in Africa, including investing in a power plant in Zimbabwe and railway construction in Sudan.
- There are many benefits, but some people think it is exploiting Africa's resources in order to benefit China's own economy.

Industrial development

Industrial development brings employment, higher incomes and opportunities to invest in housing, education and *infrastructure* (e.g. transport networks, water and sewerage). The population then becomes better educated and healthier, which provides more opportunities to invest in industries and business. This circular process is called the *multiplier effect*.

Industrial development in Malaysia

- Malaysia has seen a dramatic growth in its wealth since the 1970s.
- It has made use of foreign investment to exploit its natural resources and develop a thriving manufacturing sector.
- Today, Malaysia has a highly-developed mixed economy.

Sector	Average % annual growth rate 2011–15	% share of GDP in 2015
Services	7.2	58.3
Manufacturing	5.7	26.3
Construction	3.7	2.9
Agriculture	3.3	6.6
Mining	1.1	5.9

Figure 1 *Malaysia's economic profile*

Tourism

- Countries with tropical beaches, spectacular landscapes or abundant wildlife have become tourist destinations.
- This has led to investment and increased income from abroad, which can be used for improved education, infrastructure and housing.
- Tourism can generate a lot of income but is vulnerable in times of economic recession.

Figure 2 *A tourist village in the Seychelles*

Add a WOW! factor

Manipulate the data when describing a table of data (e.g. services had *more than twice* the percentage share of GDP in 2015 than manufacturing).

Six Second Summary

- Investment provides employment and income from abroad.
- Industrial development brings employment, higher incomes and opportunities to invest.
- Tourism has led to investment and increased income from abroad.

Over to you

Explain **three** ways in which investment can reduce the development gap.

• how aid and intermediate technology can reduce the development gap.

*Student Book
See pages 210–11*

What is aid?

Aid is when a country or non-governmental organisation (NGO), such as Oxfam, donates resources to another country to help it develop or improve people's lives. In 2013, Pakistan received £338 million from the UK. It was spent mainly on education and to reduce hunger and poverty.

Aid can take the form of:

• money
• emergency supplies
• food or technology
• specialist skills (i.e. doctors or engineers).

Aid can reduce the development gap by:

• enabling countries to invest in development projects such as roads
• focusing on health care, education and services at a local scale.
• Only aid that is long-term and freely given can really address the development gap.

Goat Aid from Oxfam

Goat Aid Oxfam was set up to help African families to buy a goat, which produces milk (which can be used to make butter), and meat. This helps to generate food, fertiliser and income, and builds community spirit.

Intermediate technology

Intermediate technology is sustainable and appropriate to the needs, knowledge and wealth of local people. It takes the form of small-scale projects.

One such project is at Adis Nifas in Ethiopia where a small dam was built, creating a reservoir close to fields for irrigation.

The benefits of using intermediate technology at the Adis Nifas dam:

• used (and still uses) intermediate technology to build and run a small dam
• used local building materials
• provided local employment
• used local tools and knowledge
• the irrigated land provides food for the villagers.

 Six Second Summary

• Aid can be invested in development projects or local-scale projects.
• Intermediate technology involves local people and is appropriate to their needs.

 Over to you

Record your understanding of aid and intermediate technology on a voice recorder so that you can listen to it again.

Reducing the gap – fair trade

You need to know:

- how fair trade can reduce the development gap.

Student Book
See pages 212–13

Is trade fair?

Rich countries protect their **trade** using two main systems.

1 *Tariffs* are taxes paid on imports, making imported goods more expensive and locally produced goods more attractive.

2 *Quotas* are limits on the quantity of goods, usually primary products, that can be imported.

Cocoa from Ghana

The EU charges 15% import tariff on chocolate, but no tariff on raw cocoa beans. Therefore, Ghana is forced to export cocoa beans rather than develop its own industry of making chocolate, which would be more valuable.

Figure 1 *Cocoa farmer in Ghana*

What is free trade?

Free trade is when countries do not charge tariffs and have quotas. This has the potential to benefit the world's poorest countries.

Subsidies are a barrier to free trade. Rich countries can afford to pay subsidies to farmers, so their products are cheaper than those produced by poorer countries.

There are advantages of joining a *trading group* for poor countries:

- it encourages free trade between members

- members are able to get higher prices for their goods.

What is Fairtrade?

Fairtrade is an international movement that sets standards for trade and helps to ensure that producers in poor countries get a fair deal.

The farmer gets a fair price and all the money from the sale of the crop.

Farming is done in an environmentally friendly way.

Part of the price is invested in local community development projects.

The product gains a stronger position in the global market.

FAIRTRADE INTERNATIONAL

Six Second Summary

- Free trade is when countries do not charge tariffs and quotas.
- Fairtrade helps to ensure that producers in poor countries get a fair deal.

Over to you

Create a mind map of **six** different methods that can be used to reduce the development gap.

You need to know:

- how debt relief and microfinance loans can help reduce the development gap.

Student Book see pages 214–15

How have poor countries built up debt?

Many poor countries borrowed money to develop their economies and export more. This had led to **debt crisis**. But:

- low commodity prices reduced the value of their exports
- high oil prices increased the price of imports

which resulted in:

- interest rates rising and their debt increased.

How can debt relief reduce the development gap?

- **Debt relief** has helped poor countries invest in development projects, such as infrastructure.
- Countries have used the money saved to provide services such as free education.

However, corrupt governments may keep the money for themselves. You can find out which of the world's governments are corrupt by searching for 'The world's most corrupt countries' in Google images.

What is microfinance?

- Microfinance is small-scale financial support.
- It is available from banks that are set up to help the poor.
- **Microfinance loans** enable individuals or families to start up small businesses.

The Grameen Bank in Bangladesh lends US$200 to village women to buy a mobile phone (Figure **1**). Other villagers then pay the women to use the phones.

Figure 1 *Using the village phone*

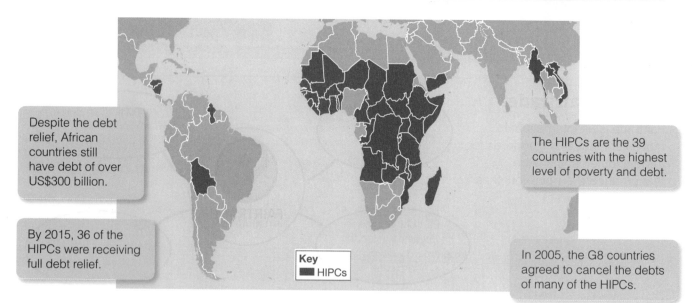

Despite the debt relief, African countries still have debt of over US$300 billion.

By 2015, 36 of the HIPCs were receiving full debt relief.

The HIPCs are the 39 countries with the highest level of poverty and debt.

In 2005, the G8 countries agreed to cancel the debts of many of the HIPCs.

Key HIPCs

Figure 2 *The highly indebted poor countries (HIPCs) in 2016*

Six Second Summary

- Debt relief can help poor countries invest in development projects.
- Microfinance is small-scale financial support to help people start up small businesses.

Over to you

- Write a mini-test about ways of reducing the development gap.
- In a few days' time, see if you can answer the questions.

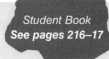
Student Book
See pages 216–17

You need to know:

- how tourism in Jamaica can help reduce the development gap
- that Jamaica is an example of the growth of tourism in an LIC or NEE.

How has tourism contributed to Jamaica's development

Over the last few decades, tourism has helped raise the level of development in Jamaica and reduce the development gap.

 **Big Idea**

Jamaica is an example of the growth of tourism in an LIC or NEE.

Example

Economy

- In 2014, tourism contributed 24% of Jamaica's GDP.
- Income from tourism is US$2 billion each year and taxes contribute further to the development of the country.

Infrastructure

- Tourism has led to a high level of investment on the north coast.
- Improvements in roads and airports have been slower than other facilities.
- Some parts of the island remain isolated.

Quality of life

- In the northern tourist areas of Montego Bay and Ocho Rios, wealthy Jamaicans have a high standard of living (Figure **2**).
- However, large numbers of people live nearby in poor housing with inadequate access to fresh water, health care and education.

The environment

- Conservation and landscaping projects provide job opportunities.
- Community tourism and sustainable *ecotourism* is expanding in more isolated regions.
- Mass tourism can create environmental problems such as footpath erosion, excessive waste and harmful emissions.

Employment

- Tourism provides 200 000 jobs, either directly or indirectly.
- Employment in tourism provides income which helps to boost the local economy.
- Those in employment learn new skills.

Figure 1 *Tourism boosts the local economy*

Figure 2 *Turtle Beach, Ocho Rios, Jamaica*

Six Second Summary

- Tourism creates employment and investment and can boost the economy.
- Tourism does not necessarily bring benefits for everyone.

 Over to you

1 Revise the section headed 'Economy'.
2 Cover it and write down all you remember.
3 Check it against the book and note anything you have missed.
4 Repeat for the other headings.

You need to know:

- where Nigeria is located
- Nigeria's global and regional importance.

Student Book
See pages 218–19

Where is Nigeria?

Nigeria is in West Africa, bordering four countries. It extends from the Gulf of Guinea in the south to the Sahel in the north (Figure **2**).

Big Idea

Nigeria is a case study of a newly emerging economy (NEE) experiencing rapid economic development. Use this study wherever questions are asked about an NEE.

What is the global importance of Nigeria?

Year	Population	Annual change %	Fertility rate	Urban population %	Urban population	% of world pop
2015	182 201 962	2.71	5.74	48.10	87 680 500	2.63
1990	95 617 345	2.65	6.6	29.70	28 379 229	1.97

Figure 1 *How has Nigeria developed in 25 years?*

In 2014, Nigeria was the 21st largest economy in the world and it is still growing.

Nigeria supplies 2.7% of the world's oil. Much of the country's economic growth has been based on oil revenues.

It has developed a diverse economy, including financial services, telecommunications and media.

Nigeria is the fifth largest contributor to UN global peacekeeping missions.

Nigeria's regional importance in Africa

Nigeria has one of the fastest-growing economies in Africa.

In 2014 it had Africa's highest GDP and the third largest manufacturing sector.

It has the largest population of any African country.

Nigeria has the highest farm output in Africa. A large proportion of people are employed in agriculture; most are subsistence farmers.

Has huge potential despite problems with internal corruption and lack of infrastructure.

Figure 2 *The location of Nigeria*

Six Second Summary

- Nigeria is in West Africa.
- Nigeria is an important oil producer and is experiencing rapid economic development.
- Nigeria is important regionally in terms of its GDP, population and farm output.

Over to you

Annotate **six** features of Nigeria on a map to show its location and its importance globally and regionally. Stick it on your wall as a reminder.

You need to know:

- the political, social, cultural and environmental contexts of Nigeria.

Student Book
see pages 220–1

Political context

- During the colonial period, Europeans exploited Africa's resources and people. Nigeria was ruled by the UK as a colony. It became independent in 1960.
- Political instability affected Nigeria's development and led to widespread corruption.
- Since 1999, it has had a stable government.
- Several countries are now starting to invest in Nigeria (e.g. China, USA).

Social context

- Nigeria is a multi-ethnic, multi-faith country – this is a strength, but has also been a source of conflict including a civil war between 1967 and 1970.
- Recently, economic inequality between the Islamic north and Christian south of Nigeria has created new religious and ethnic tensions. This has created an unstable situation with a negative impact on the economy.

Cultural context

- Nigerian music – e.g. the musician Fela Kuti.
- Nigerian cinema – 'Nollywood' – is the second largest film industry in the world, behind India.
- Well-known Nigerian writers include Wole Soyinka.
- The Nigerian football team has won the African Cup of Nations three times.

Environmental context

Nigeria's natural environments form a series of bands because of decreasing rainfall towards the north.

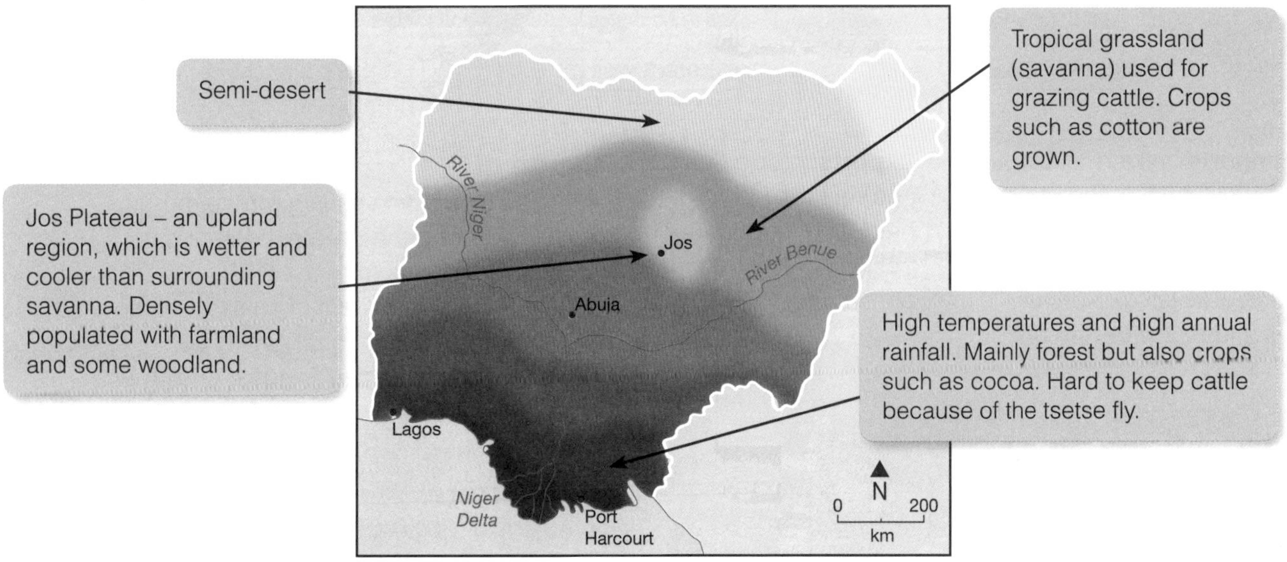

Semi-desert

Tropical grassland (savanna) used for grazing cattle. Crops such as cotton are grown.

Jos Plateau – an upland region, which is wetter and cooler than surrounding savanna. Densely populated with farmland and some woodland.

High temperatures and high annual rainfall. Mainly forest but also crops such as cocoa. Hard to keep cattle because of the tsetse fly.

River Niger

River Benue

Jos

Abuja

Lagos

Niger Delta

Port Harcourt

0   N   200
km

Figure 1 *Nigeria's natural environment*

Six Second Summary

- There is considerable variety in Nigeria socially, culturally and environmentally.
- There have been political struggles but recently the government has been stable.

Over to you

Pick out **five** important pieces of information using sticky notes or by highlighting text on this page. Learn this information, and use the highlighting and sticky notes to jog your memory later.

Student Book
See pages 222–3

> **You need to know:**
> - about Nigeria's changing relationships with the wider world.

How have Nigeria's political links changed?

- Until 1960, Nigeria was part of the British Empire.
- Since independence, Nigeria has become a member of the British **Commonwealth**.
- Nigeria is also a leading member of African political and economic groups, and international organisations.

OPEC (Organisation of Petroleum Exporting Countries): aims to stabilise the price of oil and to ensure a regular supply

ECOWAS (Economic Community of West African States): trading group

Nigeria's political links

CEN-SAD (Community of Sahel-Saharan States): trading group and develops sporting links

United Nations: Nigeria has a significant role in peacekeeping

African Union: economic planning and peacekeeping group

What are Nigeria's global trading relationships?

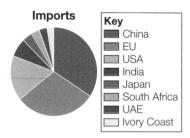

Imports

Key
- China
- EU
- USA
- India
- Japan
- South Africa
- UAE
- Ivory Coast

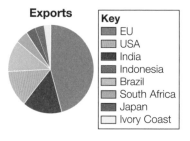

Exports

Key
- EU
- USA
- India
- Indonesia
- Brazil
- South Africa
- Japan
- Ivory Coast

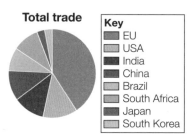

Total trade

Key
- EU
- USA
- India
- China
- Brazil
- South Africa
- Japan
- South Korea

Figure 1
Nigeria's trading relationships

Imports: refined petroleum from the EU and the USA; cars from Brazil and the USA. Telephones from China is a fast-growing import.

Exports: crude and refined petroleum, natural gas, rubber, cocoa and cotton.

Crude oil
- Crude oil dominates Nigeria's exports.
- Until recently, the greatest demand for Nigerian oil was from the USA.
- With the development of shale oil in the USA, demand for Nigerian oil has fallen.
- India is now Nigeria's biggest customer.

> ⏱ **Six Second Summary**
>
> - Nigeria's political role used to be focused on the British Empire.
> - It is now a member of African and international groups.
> - The EU is Nigeria's main trading partner.

> ✏ **Over to you**
>
> Write down **eight** pieces of information you've learnt from this page.

Agriculture
- Australia (30%) and Indonesia (15%) are the biggest customers for Nigerian cotton.
- Only two other West African countries are significant trading partners – Ghana and Ivory Coast.

Figure 2 *Picking cotton for export*

*Student Book
See pages 224–5*

You need to know:

- how Nigeria's economy is changing.

Nigeria's sources of income

Traditionally, primary products such as cocoa and cotton were Nigeria's main source of income. Today, oil accounts for 95% of Nigeria's export earnings.

 Big Idea

Nigeria's **industrial structure** (the proportion of the workforce employed in different sectors) is changing.

Does Nigeria have a balanced economy?

Changes in Nigeria's economy since 1999 are outlined in Figure **1**. This means that Nigeria now has a more *balanced* economy.

The growth of communications, retail and finance in the service (*tertiary*) sector.

Employment in agriculture (*primary sector*) has fallen, due to increasing use of farm machinery and better pay and conditions elsewhere.

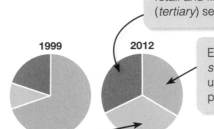

1999   2012

Key
- Agriculture
- Industry
- Services

Industrialisation and economic growth has increased employment in oil production, manufacturing and other industries (*secondary sector*).

Figure 1 *Changes in Nigeria's employment structure, 1999–2012*

Nigeria's growing manufacturing sector

- Manufacturing involves making products from raw materials.
- Manufacturing growth has been hindered by the dependence on the export of raw materials.
- Today, manufacturing accounts for 10% of Nigeria's GDP.

How is manufacturing affecting economic development?

- Regular paid work gives people a more secure *income*. This means there is a *large home market* for products manufactured in Nigeria, such as cars.
- Manufacturing industries *stimulate growth for other companies*, such as those supplying parts to make cars.
- More people are employed, so *revenue from taxes increases*.
- A thriving industrial sector attracts *foreign investment*.
- Oil processing has led to the *growth of chemical industries*, including soaps and plastics.
- This growth is an example of the *multiplier effect* (see 16.8).

Figure 2 *Volkswagen car factory in Lagos*

 Six Second Summary

- Nigeria's economy has become more balanced.
- Manufacturing can increase incomes, attract investment and encourage the growth of linked industries – the multiplier effect.

Over to you

- Add doodles to this page to help you remember the content – e.g. a tractor for primary sector.
- Draw a flow diagram to show Nigeria's economic multiplier effect.

Student Book
See pages 226–7

You need to know:

• the role of TNCs in Nigeria's development.

A **transnational corporation (TNC)** is a large company that operates in several countries.

A TNC usually has its headquarters in one country with production plants in several others.

What are the advantages and disadvantages of TNCs in Nigeria?

Advantages

😀 Companies provide employment and the development of new skills.

😀 Investment by companies in local infrastructure and education.

😀 Other local companies benefit from increased orders.

😀 Valuable export revenues are earned.

Disadvantages

🙁 Local workers are sometimes poorly paid.

🙁 Working conditions are sometimes very poor.

🙁 Management jobs often go to foreign employees.

🙁 Much of the profit goes abroad.

Unilever in Nigeria

Unilever is an Anglo-Dutch TNC, manufacturing items such as soap, foods and personal care items.

• Unilever employs about 1500 people in Nigeria.
• It has promoted improvements in health care, education and water supply.

Figure 1 Unilever's Agbara factory in Ogun State

Shell Oil in the Niger Delta

Shell is one of the world's largest oil companies. It has extracted oil from the Niger Delta since 1958, with some controversy.

Advantages

😀 It has made major contributions in taxes.

😀 It has provided direct employment for 65 000 Nigerian workers.

😀 It has provided 250 000 jobs in related industries.

😀 91% of all Shell contracts have been placed with Nigerian companies.

Disadvantages

🙁 Oil spills have caused water pollution and soil degradation, damaging agriculture and fishing industries.

🙁 Frequent oil flares send toxic fumes into the air.

🙁 Oil theft and sabotage cost TNCs and the government billions of dollars every year.

 **Six Second Summary**

• Advantages of TNCs include employment and investment.
• Disadvantages include profit going abroad and environmental problems.

Over to you

Think of a favourite song and change the lyrics to remind you of the advantages and disadvantages of TNCs.

*Student Book
See pages 228–9*

You need to know:

- the impact of international aid on Nigeria.

Types of aid

Aid can be provided by individuals, charities, NGOs, governments and international organisations. There are two main types of aid:

- *Emergency aid* – following a natural disaster or conflict
- *Developmental aid* – long-term support aimed at improving quality of life

Types of aid

Short-term – emergency help usually in response to a natural disaster

Long-term – sustainable aid that seeks to improve resilience, e.g. wells to reduce the effects of drought

Bilateral – aid from one country to another (which is often tied)

Tied – aid may be given with certain conditions

Multilateral – richer governments give money to an international organisation, e.g. the World Bank, which then redistributes to poorer countries

Voluntary – money donated by the public and distributed by NGOs such as Oxfam

What is the impact of aid in Nigeria?

The most successful aid projects are community-based, supported by small charities and NGOs. Aid has been used to benefit Nigeria in several ways.

How does aid benefit Nigeria?

In 2014, the World Bank approved a US$500 million loan to fund development projects and provide loans to businesses.

Aid from the USA helps to protect people against the spread of AIDS/HIV.

The Community Care in Nigeria project provides support for orphans.

Nets for Life (an NGO) provides education on malaria prevention and distributes anti-mosquito nets.

What prevents aid from being used effectively?

- Corruption is a major factor in loss of aid.
- Donors may have political influence over what happens to aid.
- By receiving aid, a country may become more dependent.

 Six Second Summary

- There are several different types of aid.
- Impacts of aid vary according to whether or not it is used effectively.

 Over to you

Try this 4-mark question.

Write down **two** reasons why aid may not always be effective. Explain your reasons fully.

Student Book
See pages 230–1

You need to know:

- the environmental impacts of economic development in Nigeria.

The effect of economic growth on the environment

Rapid economic growth, like in Nigeria, can bring many benefits. But it can also have a negative impact on the environment.

Industrial growth

- In Kano and Lagos, industrial pollutants go directly into water channels. They are harmful to people and ecosystems.
- Industry emits poisonous gases that can cause respiratory and heart problems (Figure **1**).
- 70–80% of Nigeria's forests have been destroyed through factors such as agriculture, urban expansion and industrial development.

Figure 1 *Air pollution in Lagos*

Commercial farming and deforestation

- There is water pollution due to chemicals, soil erosion and silting of river channels.
- Many species (including cheetahs and giraffes) have disappeared because of deforestation.

Urban growth

- Waste disposal is a major issue (Figure **2**).
- Traffic congestion leads to high levels of exhaust emissions.
- The development of Abuja has resulted in areas of rich natural vegetation being replaced by concrete.

Figure 2 *Rubbish dumped on the roadside*

Mining and oil extraction

- Tin mining led to soil erosion. Local water supplies were polluted with toxic chemicals.
- Oil spills can cause fires, sending CO_2 and other harmful gases into the atmosphere, creating *acid rain*.

Bodo oil spills (2008–09)

In 2008 and 2009, 11 million gallons of crude oil spilt over a 20 km² area around Bodo in the Niger Delta. This had disastrous impacts on the ecosystem and devastated the livelihoods of local farmers and fishermen. In 2015, Shell agreed to pay £55 million compensation to individuals and the community.

Figure 3 *An oil-polluted fish farm in Bodo*

Six Second Summary

Environmental impacts of economic development include:

- air pollution
- damage to ecosystems
- oil spills.

Over to you

Design a poster which shows your understanding of **three** environmental impacts of economic development.

Student Book
See pages 232–3

You need to know:

- how economic development has affected the quality of life for people in Nigeria.

Quality of life

As a country's economy develops, the quality of life of ordinary people should improve.

Higher disposable income to spend (e.g. on schooling)

Improvements to infrastructure, such as roads

Better access to safe water and sanitation

Improved access to a better diet means higher productivity

Better-quality health care

Reliable electricity supplies

Reliable, better-paid jobs in manufacturing or services

Figure 1 *The benefits of economic development*

Have all Nigerians benefited from economic development?

Nigeria's HDI has been increasing steadily since 2005 and is expected to continue to rise. Most indicators suggest that economic development since 1990 has improved the quality of people's lives.

Has it all been good news?

- Many people in Nigeria are still poor.
- The gap between rich and poor has become wider.
- Corruption has been a major factor and the oil wealth was not used to diversify the economy.
- Nigeria's over-dependence on oil could become a problem in the future.

Will people's quality of life continue to improve?

Sixty per cent of Nigerians live in poverty. If their quality of life is to be improved the following challenges must be met.

Political – there is a need for a continuing stable government to encourage inward investment.

Environmental – there are threats of disease spread by the tsetse fly, desertification and pollution by oil spills.

Social – historical distrust remains between tribal groups. Kidnappings by the militant group Boko Haram spread fear among Nigerians and potential investors.

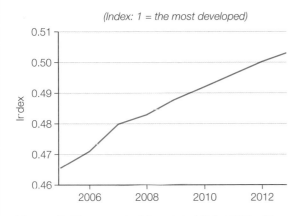

(Index: 1 = the most developed)

Figure 2 *Changes in Nigeria's HDI, 2000–13*

Six Second Summary

- Quality of life has improved as a result of economic development.
- There are still challenges to overcome as most of the population still live in poverty.

Over to you

Meet with or message a friend and take it in turns to tell each other what you have learnt about Nigeria. Continue until you both run out of ideas.

You need to know:

- how the economy of the UK has changed
- the causes of economic change in the UK.

Student Book
See pages 234–5

How has the economy of the UK changed?

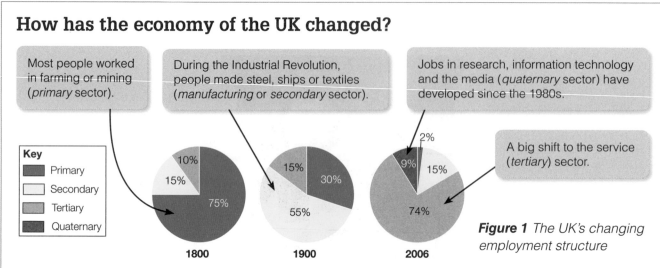

Most people worked in farming or mining (*primary* sector).

During the Industrial Revolution, people made steel, ships or textiles (*manufacturing* or *secondary* sector).

Jobs in research, information technology and the media (*quaternary* sector) have developed since the 1980s.

A big shift to the service (*tertiary*) sector.

Key
- Primary
- Secondary
- Tertiary
- Quaternary

1800 — 10%, 15%, 75%

1900 — 15%, 30%, 55%

2006 — 2%, 9%, 15%, 74%

Figure 1 *The UK's changing employment structure*

What are the causes of economic change in the UK?

De-industrialisation is the decline traditional industries, such as manufacturing. This has happened because:

- machines and technology have replaced many people
- other countries (e.g. China) can produce cheaper goods because labour is less expensive.

Globalisation is the growth and spread of ideas around the world.

- Many people now work on global brands in the quaternary sector, e.g. in IT.
- Increased world trade and cheaper imported products have contributed to the decline in UK manufacturing.

Government policies

1945–79
- The government created state-run industries such as British Rail.
- Government money 'propped up' unprofitable industries.

1979–2010
- State-run industries sold to private shareholders. This is called *privatisation*.
- Many older industries closed down.
- New private companies brought innovation and change.

2010 onwards
'Rebalancing' the economy by relying less on service industries. Policies have included:

- improvements to transport (e.g. HS2)
- more investment in manufacturing
- encouraging global firms to locate in UK.

 Six Second Summary

- The UK economy has changed over time.
- De-industrialisation is the decline of manufacturing.
- Globalisation has increased the quaternary sector.
- Government policies have caused economic change.

Over to you

Make your own glossary of the key words (in bold and in italics) in this section.

A post-industrial economy

Student Book *See pages 236–7*

You need to know:

- what a post-industrial economy is
- how the development of information and technology, service industries, finance and research have moved the UK towards a post-industrial economy.

What is a post-industrial economy?

A **post-industrial economy** is where manufacturing industry declines and is replaced by growth in the service and quaternary sectors. This happened in the UK from the 1970s.

Development of information technology

The use of **information technology (IT)** is a key factor in the UK's move to a post-industrial economy.

- Internet access enables people to work from home.
- Over 1.3 million people work in the IT sector.
- The UK is one of the world's leading digital economies.

Service industries and finance

The UK service sector (including quaternary employment) has grown rapidly since the 1970s. Today it contributes over 79% of the UK's GDP.

- Finance is an important part of the service sector.
- The UK is the world's leading centre for financial services.
- The financial services sector accounts for about 10% of the UK's GDP.

Research

The UK research sector (part of the quaternary sector) employs over 60000 highly qualified people and is estimated to contribute over £3 billion to the UK economy. This sector is likely to be one of the UK economy's main growth areas in the future.

Business and financial companies

NHS

Universities

Environment Agency

BBC

Charities

Engineering

Pharmaceutical

Figure 1 *Some UK research organisations*

British Antarctic Survey

The British Antarctic Survey (BAS) employs over 500 highly skilled people in Cambridge (UK), Antarctica and the Arctic. It is linked to the University of Cambridge and helps us understand the impact of humans on the Earth's natural systems.

Figure 2 *The Rothera Research Station in Antarctica*

 **Six Second Summary**

- Developments in IT have encouraged the growth of service and quaternary sectors.
- Finance is an important part of the service sector.
- Research employs highly qualified people.

 **Over to you**

Explain **three** reasons why the UK is developing as a post-industrial economy.

*Student Book
See pages 238–9*

You need to know:

- what a science park is
- what a business park is
- how science and business parks are moving the UK towards a post-industrial economy.

What is a science park?

A **science park** is a group of scientific and technical knowledge-based businesses located on a single site. Most are associated with universities, enabling them to use research facilities and employ skilled graduates. Science parks may also include support services such as financial services and marketing.

University of Southampton Science Park

Southampton Science Park includes one hundred small science and innovation businesses including Fibrecore (manufacturer of optical fibres) and PhotonStar (specialising in lighting products).

Benefits

- Excellent transport links – close to M3, Southampton international airport and rail links
- Excellent links with the University
- Attractive location with green areas

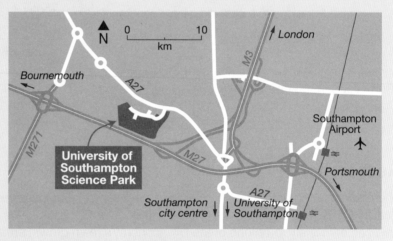

Figure 1 Southampton Science Park – transport links

What is a business park?

A **business park** is an area of land occupied by a cluster of businesses. Business parks are usually located on the edges of towns where:

- land is cheaper and more available
- access is better with less congestion
- businesses can benefit from working together.

Cobalt Business Park, Newcastle-upon-Tyne

Cobalt Park is the UK's largest business park, with support facilities including retail outlets and a fitness centre. The park is next to the A19, close to the A1 and 20 minutes from the international airport.

Businesses locating in Cobalt Park qualify for governmental assistance.

Companies in the park include Siemens, IBM and Santander.

Figure 2 Siemens offices at Cobalt Business Park

Six Second Summary

- Science parks provide benefits such as links with universities and attractive locations.
- Business parks tend to be located on the edges of towns because land is cheaper, has better access and has potential for expansion.

Over to you

Cover up everything apart from Figure 1. Use the map to explain the benefits of science parks.

Environmental impacts of industry

Student Book
see pages 240–1

- what the impacts of industry are on the physical environment
- about an example of how modern industrial development can be made more environmentally sustainable – Torr Quarry in Somerset.

Example

Impacts of industry on the physical environment

- Manufacturing plants can look dull and affect the visual effect of the landscape.
- Industrial processes and waste products can cause air, water and soil pollution.
- The transport of raw materials and manufacturing products increases levels of air pollution.

How can industrial development be more sustainable?

- Care in design can reduce the visual impact.
- Technology can be used to reduce harmful emissions.
- Desulphurisation can remove harmful gases.
- Heavy fines can be imposed when pollution incidents occur.

Quarrying in the UK

Impacts of quarrying

- destroy natural habitats
- pollute water courses
- scar landscapes

Making quarrying more sustainable

- There are strict controls on blasting, removal of dust from roads and landscaping.
- Recycling is encouraged.
- Companies are expected to restore or improve a quarry after it has been used.

Torr Quarry, Somerset

Torr Quarry is a limestone quarry in the Mendip Hills. It employs over 100 people and contributes more than £15 million towards the local economy each year. Torr Quarry is an example of how modern industrial development can be more environmentally sustainable.

- The quarry is being restored to create wildlife lakes.
- 200 acres of the site have already been landscaped.
- Regular monitoring of noise, vibration, dust and **water quality**.
- Rail transport of quarried rock minimises the impact on local roads and villages.

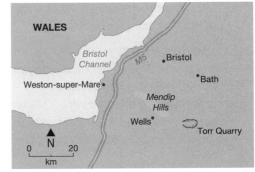

Figure 1 *Location of Torr Quarry, Somerset*

Figure 2 *Torr Quarry*

Figure 3 *Planned restoration of Torr Quarry*

Six Second Summary

- Modern industry can cause pollution and destroy habitats.
- Torr Quarry aims to be environmentally sustainable by landscaping and monitoring.

Over to you

Draw a field sketch of Figure **3** and annotate it to show how Torr Quarry is environmentally sustainable.

Student Book
See pages 242–3

You need to know:

- what social and economic changes are happening in an area of population growth and also in one of population decline.

An area of population growth: South Cambridgeshire

What are the changes?

- The population of 150 000 is increasing, due to migration into the area.
- Most migrants come from Cambridge and other parts of the UK; many arrive from Eastern Europe.
- The proportion of people aged 65 or over is growing.

Social effects

- 80% car ownership leads to increased traffic on narrow roads.
- Housing developments on the edges of villages can lead to a reduction in community spirit.
- Young people cannot afford the high cost of houses and move away.

Economic effects

- A reduction in agricultural employment as farmers sell land for housing.
- Lack of affordable housing.
- High demand leads to high petrol prices.
- Increased population puts pressure on services.

Figure 1 *Location of South Cambridgeshire and the Outer Hebrides*

Figure 2 *The landscape of Cambridgeshire*

An area of population decline: the Outer Hebrides

What are the changes?

- The population has declined by more than 50% since 1901.
- With limited employment, young people have moved away.

Social effects

- The expected fall in the number of children may result in school closures.
- An increasingly ageing population has fewer young people to support them.

Economic effects

- Services are closing.
- Most small farms (crofts) can only provide work for two days a week.
- There has been an increase in tourism, but ...
- ... the current infrastructure cannot support the scale of tourism needed to provide an alternative source of income.

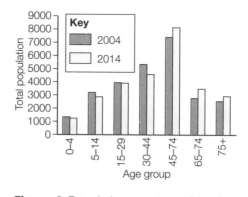

Figure 3 *Population structure of the Outer Hebrides, 2004–14*

Over to you

Draw a table comparing the populations of Cambridgeshire and the Outer Hebrides.

Changing transport infrastructure (1)

Student Book
See pages 244–5

You need to know:

- about improvements and new developments in road and rail infrastructure in the UK
- how those improvements will make a difference.

Road improvements

The 2014 'Road Investment Strategy' includes:

- 100 new road schemes by 2020
- 1300 new lane miles added to motorways and trunk roads
- extra lanes added to turn main motorways into 'smart motorways'.

New road schemes will create thousands of construction jobs and boost local and regional economies.

South-west 'super highway'

- A £2 billion road-widening project will take place on the A303.
- Converting the route to dual carriageway will create a 'super highway' to Exeter and beyond.

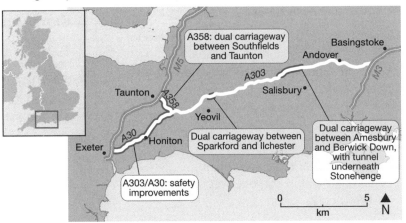

Figure 1 *Upgrading the A303*

Railway improvements

There are plans to stimulate economic growth in the north of the UK by:

- improving trans-Pennine rail links reducing journey times by up to 15 minutes.
- HS2 – a planned high-speed rail line to connect London with Birmingham, Sheffield, Leeds and Manchester. It is controversial as the route passes through countryside and close to many homes.

London's Crossrail

Crossrail is a new railway across London that links Reading and Heathrow (to the west), to Shenfield and Abbey Wood (to the east).

- Crossrail (and Crossrail 2) will reduce journey times across London.
- It will bring an additional 1.5 million people within 45 minutes' commuting distance of London's key business districts.

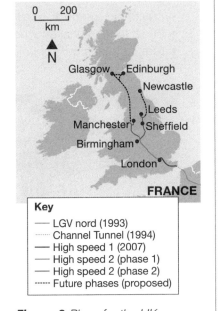

Figure 2 *Plans for the UK high-speed rail*

You need to know:

- how new developments will affect the UK's port and airport capacity
- how these can make a difference in the UK.

*Student Book
See pages 246–7*

Developing the UK's ports

About 32 million passengers travel through UK ports each year. Ports employ around 120 000 people. Private companies run many ports and invest heavily in infrastructure, often with government assistance.

- Bristol (Avonmouth) – £195 million invested for bulk handling and storage facilities.
- A new rail terminal at Felixstowe and upgraded cruise service facilities at Harwich.

Liverpool2

A new container terminal is being constructed at the Port of Liverpool, known as 'Liverpool2'.

The project will more than double the port's capacity to over 1.5 million containers a year. Phase 1 opened in 2016. When complete the new terminal will:

- create thousands of jobs in the north-west
- boost the regional economy
- reduce the amount of freight traffic on the roads.

Figure 1 *The construction of Liverpool2*

Airport developments

- Airports create vital global links
- They provide thousands of jobs
- They also boost economic growth both regionally and nationally.

Expanding London's airports

In 2015, a government report recommended a new third runway at Heathrow. The cost would be £18.6 billion. The report recommended financial support for soundproofing homes and schools, and a ban on night-time flights.

😊 This was predicted to create more jobs and make more money for the UK.

🙁 People living nearby are concerned about the noise and air pollution from planes.

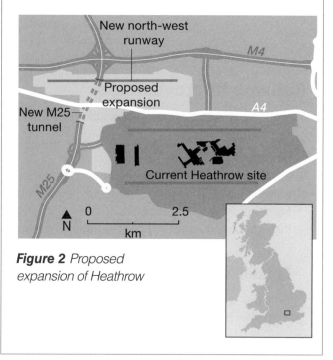

Figure 2 *Proposed expansion of Heathrow*

 Six Second Summary

- Examples of improvements and new developments are Liverpool2 (a port) and a new runway at Heathrow (an airport).
- They are intended to create new jobs and boost the economy.

 Over to you

Produce a **five**-sentence summary of what you have learnt from this page.

136 **Chapter 18** – The changing UK economy

The north–south divide

*Student Book
See pages 248–9*

What is the north–south divide?

It refers to real or imagined cultural and economic differences between the south of England and the rest of the UK.

In general, the south enjoys higher incomes and longer life expectancy. But the south also has higher house prices and more traffic congestion.

Why is there a north–south divide in the UK?

- During the Industrial Revolution, the UK's growth was centred on coalfields, heavy industries and engineering in northern England, Wales and Scotland.

- Since the 1970s, many industries have declined, reducing prosperity in those areas.

- London and the South East developed rapidly due to a fast-growing service sector.

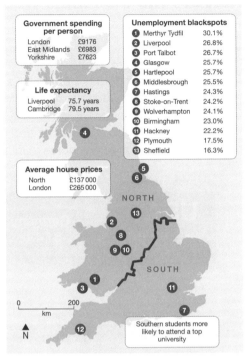

Figure 1 North and south – some facts

How can regional strategies address the issue?

Local enterprise partnerships (LEPs)

LEPs are voluntary partnerships between local authorities and businesses.

Their aim is to identify business needs and encourage companies to invest in order to boost the local economy and create jobs.

The Lancashire LEP is one example.

Lancashire LEP

The Lancashire LEP will:

- promote new businesses and create 50 000 new jobs by 2023
- improve transport with £20 million investment
- extend superfast broadband across 97% of the region
- create 6000 high-skilled jobs in Enterprise Zones at Samlesbury and Warton.

Enterprise Zones

The aim of Enterprise Zones is to encourage new businesses and jobs. The government supports businesses in Enterprise Zones by:

- providing a business rate discount
- ensuring the provision of superfast broadband
- creating simpler planning regulations.

 Six Second Summary

- The north–south divide refers to differences between the north and south of the UK.
- Local enterprise partnerships and Enterprise Zones are attempts to resolve regional differences.

Over to you

Write the words 'north-south divide' and, for each letter, write a word or phrase, relevant to this page's information, e.g. **N**orth–south divide, regi**O**nal strategies, Ente**R**prise Zones etc.

The UK in the wider world (1)

Student Book
See pages 250–1

You need to know:

- what links exist between the UK and the wider world, through trade, culture, transport and electronic communication.

What are the UK's links with the wider world?

Trade

- The UK's most important trading links are with the EU. Goods can be traded between member states without tariffs. This may change when the UK leaves the EU.
- The USA is an important historic trading partner.
- There has been a recent increase in trade with China.

UK export destinations (£ millions)

USA **38 858**	Netherlands 22 553
	Germany **31 432**
	France 19 627
	Switzerland 21 313
China **15 934**	Italy 8708
	Spain 8774
	Rep. of Ireland 18 124
	Belgium 12 597

Sources for UK imports (£ millions)

Germany **59 365**	Netherlands 31 068
	USA **32 863**
	Italy 16 663
Norway 14 958	France **24 906**
Spain 13 031	Rep. of Ireland 11 695
	Belgium **20 585**
	China **33 891**

Figure 1 *The UK's main trading partners*

Transport

- London Heathrow is one of the busiest airports in the world.
- There are important transport links between the UK and mainland Europe via the Channel Tunnel and sea ferries.

Electronic communication

- 99% of internet traffic passes along a network of submarine high-power cables.
- Connections are concentrated between the UK and USA.
- There is a further concentration in the Far East.
- A project known as Arctic Fibre is due to connect London and Tokyo.

Culture

- The global importance of the English language has given the UK strong cultural links with many parts of the world.
- Music, books and films from the UK are accessed all over the world.
- Migrants have brought their own culture to the UK, such as food and festivals.

Television

Television is one of the UK's most successful media exports. In 2013–14 it accounted for over £1.28 billion of export earnings.

Figure 2 Dr Who – *a UK export success*

 Six Second Summary

- The UK is connected to the wider world via trade, culture, transport and electronic communication.
- These links often generate more money for the UK.

Over to you

Talk for one minute about the links the UK has with the rest of the world.

Student Book
See pages 252–3

You need to know:

- about the UK's economic and political links with the European Union (EU) and the Commonwealth.

What are the UK's links with the European Union?

The EU has 28 member countries. It is an important trading group, but its powers also include political influence.

Many in the UK feel that the EU is too influential in making laws which affect the UK. In June 2016, the people of the UK voted to leave the EU. The UK will remain a member until exit negotiations are completed.

Financial support for farmers and disadvantaged regions in the UK.

How has the EU affected the UK?

Goods, services, capital and labour can move freely between member states and encourage trade.

There are EU laws and controls on crime, pollution and consumers' rights.

In 2013, about 40% of total UK immigrants were from the EU.

What are the UK's links with the Commonwealth?

The UK is a member of the Commonwealth, a voluntary group of countries, most of which were once British colonies.

The Commonwealth Secretariat provides advice and support to member countries on a range of issues including human rights and social and economic development.

There are important trading and cultural links between the UK and the Commonwealth countries. There are also sporting connections, such as the Commonwealth Games.

Figure 1 *The Commonwealth Games in Glasgow, 2014*

Six Second Summary

- The UK has both political and economic links with the EU and the Commonwealth.
- Economic links include trading links.
- Political links include laws or advice and support.

Over to you

- Highlight this page to show which links are economic and which are political.
- Create a Venn diagram to show these links.

Section C
The challenge of resource management

Your exam

Section C The challenge of resource management makes up part of Paper 2: Challenges in the human environment.

Paper 2 is a one-and-a-half hour written exam and makes up 35 per cent of your GCSE. The whole paper carries 88 marks (including 3 marks for SPaG) – questions on Section C will carry 25 marks.

You need to study resource management and one topic from food, water or energy in Section C – in your final exam you will have to answer Question 3 and one other question.

Tick these boxes to build a record of your revision

Your revision checklist

Spec key idea	Theme	1	2	3
19 Resource management				
Food, water and energy are fundamental to human development	19.1 The global distribution of resources			
The changing demand and provision of resources in the UK create opportunities and challenges	19.2 Provision of food in the UK			
	19.3 Provision of water in the UK			
	19.4 Provision of energy in the UK			
20 Food management				
Demand for food resources is rising globally but supply can be insecure, which may lead to conflict	20.1 Global food supply			
	20.2 Impact of food insecurity			
Different strategies can be used to increase food supply	20.3 Increasing food supply			
	20.4 The Indus Basin Irrigation System			
	20.5 Sustainable food production (1)			
	20.6 Sustainable food production (2)			
21 Water management				
Demand for water resources is rising globally but supply can be insecure, which may lead to conflict	21.1 Global water supply			
	21.2 The impact of water insecurity			
Different strategies can be used to increase water supply	21.3 How can water supply be increased?			
	21.4 The Lesotho Highland Water Project			
	21.5 Sustainable water supplies			
	21.6 Wakel River Basin project			
22 Energy management				
Demand for energy resources is rising globally but supply can be insecure, which may lead to conflict	22.1 Global energy supply and demand			
	22.2 Impacts of energy insecurity			
Different strategies can be used to increase energy supply	22.3 Strategies to increase energy supply			
	22.4 Gas – a non-renewable resource			
	22.5 Sustainable energy use			
	22.6 The Chambamontera micro-hydro scheme			

Student Book
See pages 256–7

You need to know:

- why food, water and energy are significant for economic and social well-being
- that resources are distributed unevenly around the world.

What is a resource?

- It is a stock or supply of something that has value or purpose.
- The most important resources are food, water and energy.
- Most HICs have plentiful resources, many of them imported.
- Many poorer countries lack resources and struggle to improve quality of life.

Big Idea

Population growth presents many challenges for **resource management**.

Food

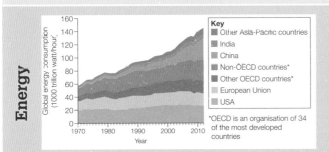

Key
% of population undernourished
- >35
- 25-34
- 15-24
- 5-14
- <5
- No data

Figure 1 *Global undernourishment*

Why are resources significant?

- A poorly balanced diet can cause illness and diseases.
- People need to be well fed to be productive.
- Obesity is an increasing problem.

What are the global inequalities?

- Over one billion people do not get enough calories.
- **Undernutrition** (malnutrition) affects a further two billion.
- Countries in sub-Saharan Africa suffer most from undernutrition.

Water

Key
- Physical water scarcity (lack of water, e.g. deserts)
- Economic water scarcity (countries that cannot afford to exploit water supplies)
- Little or no water scarcity
- No data

Figure 2 *Projected areas of water scarcity by 2025*

- Essential for drinking.
- Vital for crops.
- Used to produce energy.

- Variations in climate and rainfall affect supply.
- Capture, storage and extraction is expensive.
- Many poor countries have water shortage.
- LICs/NEEs use most water for agriculture
- HICs use most water in industry.

Energy

Key
- Other Asia-Pacific countries
- India
- China
- Non-OECD countries*
- Other OECD countries*
- European Union
- USA

*OECD is an organisation of 34 of the most developed countries

Global energy consumption (1000 trillion watt/hour)

Figure 3 *Global energy consumption*

- Needed for light, heat and power.
- Powers factories.
- Provides fuel for transport.

- Richer countries consume more energy than poorer countries.
- The Middle East is a major oil supplier; its own consumption is low.
- As NEEs become more industrialised, the demand for energy will increase.

Six Second Summary

- Richer countries use more resources than poorer countries.
- Many people in sub-Saharan Africa are undernourished and suffer from water scarcity.

Over to you

Annotate a blank world map to show the global inequalities in resource supply and consumption.

You need to know:

- how demand for food is changing in the UK
- how sourcing food affects the UK's carbon footprint
- how agribusiness works.

*Student Book
See pages 258–9*

How is demand for food changing in the UK?

The UK imports about 40% of the total food it consumes. This percentage is increasing.

> Availability of cheaper food from abroad

> Demand for more exotic foods and seasonal produce all year round

> **Why does the UK import so much food?**

> UK climate is unsuitable for production of some foods

What is the impact of importing food?

Foods can travel long distances (**food miles**). Importing food also adds to our **carbon footprint**. This comes from producing the energy for commercial cultivation, and from transport.

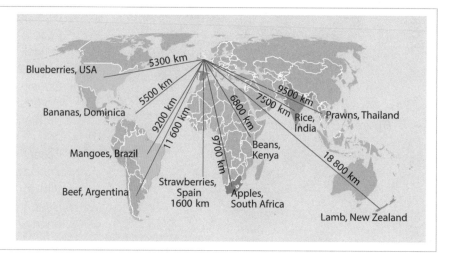

Blueberries, USA — 5300 km
Bananas, Dominica — 5500 km
Mangoes, Brazil — 9200 km
— 11 600 km
Beef, Argentina
Strawberries, Spain 1600 km — 9700 km
Apples, South Africa
Beans, Kenya — 6800 km
Rice, India — 7500 km
Prawns, Thailand — 9500 km
Lamb, New Zealand — 18 800 km

Figure 1 *Distances travelled by UK imported food*

How is the UK responding to these challenges?

People are being encouraged to eat locally produced foods according to season. Two recent trends in UK farming are agribusiness and **organic produce**.

Lynford House Farm in East Anglia – an agribusiness

- The land is intensively farmed, maximising the amount of food produced.
- Pesticides and fertilisers are widely used.
- Machinery costs are high but increase efficiency.
- A small number of workers are employed.

Riverford Organic Farms

- Began as an organic farm in Devon.
- Now delivers organic vegetables from farms in Devon, Yorkshire, Peterborough and Hampshire.
- This reduces food miles and provides local employment.

 Six Second Summary

- There is all-year demand for seasonal food and organic produce.
- Importing food increases carbon footprints.
- Agribusiness is a recent trend in UK farming.

Over to you

In two tables, list the positive and negative impacts of a) the UK's changing demand for food and b) agribusiness versus organic farming.

Provision of water in the UK

Student Book
See pages 260–1

You need to know:

- how demand for water in the UK is changing
- how supply meets demand
- how water quality and pollution are managed.

What are the demands for water in the UK?

Almost 50% of the UK's water supply is used domestically. Demand for water in the UK is estimated to rise by 5% between 2015 and 2020 because of a rapidly growing population, more houses and an increase in the use of water-intensive domestic appliances (e.g. dishwashers).

How far does the UK's water supply meet demand?

The north and west have a **water surplus**, where supply exceeds demand.

The south and east have a **water deficit**, where demand exceeds supply.

Water stress (where demand exceeds supply) is experienced in more than half of England.

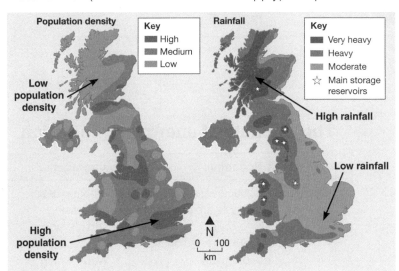

Figure 1 *UK population density and water supply*

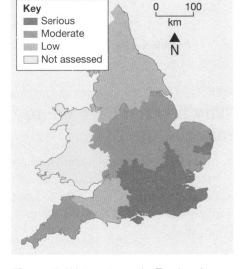

Figure 2 *Water stress in England*

Managing water quality

The Environment Agency manages water quality by:
- filtering water to remove sediment
- purifying water by adding chlorine
- imposing strict regulations.

Some groundwater sources have been polluted by:
- industrial sites discharge
- agricultural chemical fertilisers
- leaching from old underground mines.

Water transfer

There is a growing need to increase **water transfer** to meet demand.

There is opposition because of:

- the effect on land and wildlife
- high costs
- greenhouse gases released by pumping water over long distances.

Six Second Summary

- Demand for water is increasing.
- The UK has areas of water surplus and water deficit.
- Water quality and pollution need to be managed.

Over to you

Cover everything apart from the maps on this page. Use them to suggest how far the UK's water supply meets demand.

Student Book
See pages 262–3

You need to know:

- how the UK's energy mix is changing
- how there are reduced domestic supplies of coal, gas and oil
- the economic and environmental issues associated with exploiting energy sources.

How has the UK's energy mix changed?

Energy consumption has fallen in the UK in recent years, mainly due to the decline of heavy industry and energy conservation (use of low-energy appliances, building insulation, fuel-efficient cars).

By 2020, the UK aims to meet 15% of its energy requirement from renewable sources.

In 1990 almost three-quarters of UK energy came from coal and oil – 'fossil' or *non-renewable fuels*.

By 2007 there was an equal mix of coal, gas and nuclear – all non-renewable sources.

By 2014 *renewable* sources, such as wind and solar energy, had become more important.

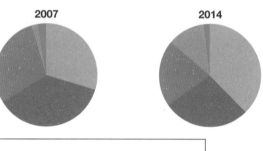

1990 **2007** **2014**

Key					
■ Coal	■ Gas	■ Renewables	■ Oil	■ Nuclear	■ Other

Figure 1 *The UK's changing energy mix*

Why has the UK's energy mix changed?

- About 75% of the UK's known oil and natural gas reserves have been used up.
- Coal consumption has declined because of concerns about greenhouse gas emissions.

The UK's **energy security** is affected as it becomes increasingly dependent on imported energy. However, fossil fuels are likely to remain important in the future because:

- the UK's remaining reserves will provide energy for several decades
- coal imports are cheap
- shale gas deposits may be exploited in the future.

The impacts of energy exploitation

	Economic	Environmental
Nuclear	• Nuclear power plants are expensive to build. • Decommissioning old plants is expensive. • New plants provide job opportunities.	• Problem of safe processing and storage of radioactive waste. • Warm waste water can harm local ecosystems.
Wind farms	• High construction costs. • Local homeowners can have lower energy bills.	• Visual impact on the landscape. • Help reduce carbon footprints. • Noise from wind turbines.

 Six Second Summary

- In the UK, fossil fuels remain important, but renewables are now significant.
- There are positive and negative economic and environmental impacts of energy source exploitation.

Over to you

Create a mini-test of **ten** questions about this whole chapter then test yourself in a few days' time.

Global food supply

Student Book
See pages 264–5

You need to know:

- how calorie intake and food supply vary around the world
- why there is increasing food consumption
- how different factors affect food supply.

Global patterns of food consumption

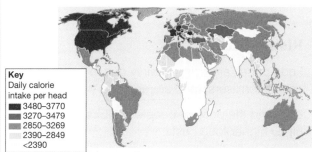

Key
Daily calorie intake per head
- 3480–3770
- 3270–3479
- 2850–3269
- 2390–2849
- <2390

Figure 1 Global food consumption

- Canada, USA and Europe consume the most calories.
- In sub-Saharan Africa, daily calorie intake per head is below the recommended daily intake of 2000–2400 calories.

Global food consumption is increasing because:

- there are growing populations
- increasing levels of development mean people can afford to buy more food
- improved transport and storage means there is more food available.

What is meant by food security?

Food security – having access to enough affordable, nutritious food to maintain a healthy life.

Countries which produce more food than is needed by their population have a *food surplus*.

Countries which do not produce enough food to feed their population and have to rely on imported food have a *food deficit*. Many of these countries also experience **food insecurity**.

Global patterns of food supply

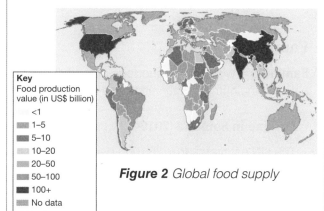

Key
Food production value (in US$ billion)
- <1
- 1–5
- 5–10
- 10–20
- 20–50
- 50–100
- 100+
- No data

Figure 2 Global food supply

- USA, Brazil and UK have high outputs due to intensive farming and investment.
- China and India have large populations and high agricultural outputs.
- Sub-Saharan African countries produce less food. They have unreliable rainfall, low investment and a lack of training.

What factors affect food supply?

- *Climate* – regions experiencing extreme temperatures and rainfall struggle to produce food.
- *Technology* – in HICs, mechanisation and agribusiness give high levels of productivity.
- *Pests and diseases* spread from the Tropics with rising temperatures.
- *Water stress* – lack of water affects many areas that suffer food scarcity.
- *Conflict* can lead to the destruction of crops and livestock.
- *Poverty* – the poorest people cannot afford technology, irrigation or fertilisers.

Six Second Summary

- Food consumption and supply vary around the world.
- Reasons for increasing food consumption include economic development and rising populations.
- Climate, technology and conflict are some factors which affect food supply.

Over to you

- Describe **four** points about global patterns of a) food consumption, b) food supply.
- Create a mind map of the factors affecting food supply.

Student Book
See pages 266–7

You need to know:

• what the impacts of food insecurity are.

What are the impacts of food insecurity?

Food insecurity occurs when a country can't supply enough food to feed its population.

Famine

Famine is a widespread shortage of food often causing malnutrition, starvation and death.

Famine in Somalia (2010–12)

The UN estimates that 258 000 people died in Somalia as a result of food insecurity during the famine of 2010-12.

The famine had two main causes:

• Two successive seasons of low rainfall, poor harvests and the death of livestock.
• In southern and central Somalia, the al-Shahab militant group blocked aid, making the crisis worse.

Rising prices

Food prices are rising, mainly due to increased cost of fertilisers, food storage and transportation.

LICs and the poorest people in NEEs are hardest hit by higher food costs.

Figure 1 *The quantities of rice the same money could buy in 2008 (left) and 2007 (right)*

Soil erosion

Soil erosion involves the removal of fertile top soil layers by wind and water. There are several causes (Figure **2**).

Figure 2 What causes soil erosion?

Overgrazing by animals reduces the amount of vegetation, leaving soil exposed.

Growing too many crops can use up valuable nutrients, reducing soil fertility.

Cultivation of marginal land to increase food production can lead to loss of fertility.

Deforestation for farming (as in the photo) removes the protective covering of the trees and increases surface run off.

Undernutrition

Undernutrition is the lack of a balanced diet, and deficiency in minerals and vitamins.

It causes around 300 000 deaths per year and contributes to half of all child deaths, particularly in southern Asia and sub-Saharan Africa.

Social unrest

Incidents of social unrest ('food riots') are often linked to large increases in the price of food.

In 2011, the price of cooking oil and flour doubled. In Algeria, this led to five days of rioting, with four people killed.

 Six Second Summary

• Impacts of food security include: famine, undernutrition, soil erosion, rising prices and social unrest.

Over to you

Create a mnemonic for the impacts of food insecurity. Add an example for each impact.

Student Book
See pages 268–9

You need to know:

- what strategies are used to increase food supply
- how each strategy works to increase food supply.

How can food supply be increased?

Irrigation

Irrigation is the artificial watering of land. Irrigation projects can involve the construction of expensive dams and reservoirs, such as in the Indus Valley of Pakistan. They often benefit larger commercial farming.

There are smaller schemes such as in Makueni County in eastern Kenya. Pipelines and storage tanks enable drip irrigation to support domestic food cultivation.

The 'new' green revolution

The 'new' **green revolution** focuses on sustainability and community. It uses techniques such as:

- water harvesting and irrigation
- soil conservation
- improving seed and livestock quality using science and technology.

Along with improved rural transport and affordable credit, these innovations have enabled the Indian state of Bahir to double its rice output.

Aeroponics and hydroponics

Aeroponics – plants are sprayed with fine water mist containing plant nutrients. Excess water is re-used. This enables small-scale farmers to increase yields and lower production costs.

Hydroponics

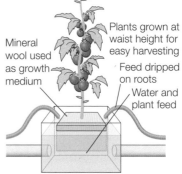

Light during winter months

Mineral wool used as growth medium

Plants grown at waist height for easy harvesting

Feed dripped on roots

Water and plant feed

Figure 1 *How hydroponics works*

Appropriate technology:

- means using skills or materials that are cheap and easily available to increase output without putting people out of work.
- is particularly appropriate for people living in poorer countries.

Figure 2 *Using a bicycle to de-husk coffee beans*

Biotechnology:

- uses living organisms to make or modify products or processes.
- includes the development of genetically modified (GM) crops, which produce higher yields and use fewer chemicals.
- In the UK, there is opposition to GM crops because of the possible effects on the environment and human health.

 Six Second Summary

- Strategies used to increase food supply include: irrigation, aeroponics and hydroponics, the new green revolution, biotechnology and appropriate technology.

 Over to you

- Make clear definitions of all the terms listed in the Six Second Summary.
- Learn these over five minutes, then write them out from memory.

The Indus Basin Irrigation System

You need to know:

- about the Indus Basin Irrigation System, a large-scale agricultural development
- that the aim of this development is to increase food supply
- the advantages and disadvantages of this development.

Student Book
See pages 270–1

The Indus River runs from the Tibetan Plateau, through Pakistan to the Arabian Sea. With its tributaries, it supplies water to irrigate the drier agricultural land further south.

What is the Indus Basin Irrigation System (IBIS)?

- The IBIS is the largest continuous irrigation scheme in the world.
- Three large dams and over a hundred smaller dams regulate water flow.
- Link canals enable water to be transferred between rivers.
- Smaller canals distribute the water across the countryside.
- Over 1.6 million km of ditches and streams provide irrigation for Pakistan's agricultural land.

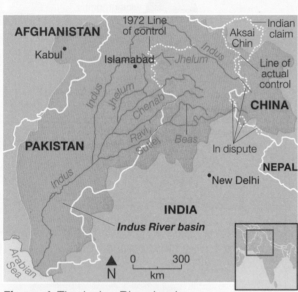

Figure 1 *The Indus River basin*

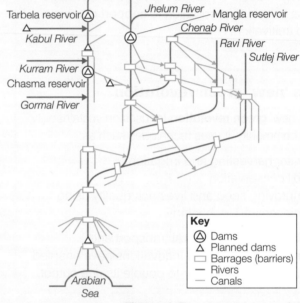

Figure 2 *Topological map of the IBIS*

What are the advantages and disadvantages of the IBIS?

Advantages

- Improves food security for Pakistan, making 40% more land available for cultivation.
- Irrigation has increased crop yields.
- Diets have improved as a greater range of food products is available.
- HEP is generated by the large dams.

Disadvantages

- Some farmers take an unfair share of water.
- Poor irrigation techniques mean water is wasted. *Salinisation* (increased saltiness) can damage the soil.
- Population growth will increase the demand for water.
- High costs to maintain reservoir capacity.

Six Second Summary

- The Indus Basin Irrigation System (IBIS) is an example of a large scale agricultural development to increase food supply.
- It has improved food security, but some farmers take an unfair share of the water.

Over to you

- Write down **two** advantages and **two** disadvantages of the IBIS.
- For each statement, explain why it benefits/does not benefit people in Pakistan.
- State which was the most significant advantage and disadvantage, and explain why.

Sustainable food production (1)

Student Book
See pages 272–3

You need to know:

- what a sustainable food supply is
- how different strategies can create the potential for sustainable food supplies.

What is sustainable food supply?

A **sustainable food supply** ensures that fertile soil, water and environmental resources are available for future generations.

Organic farming

Organic farming is growing crops or rearing livestock without the use of artificial chemicals. Many people choose to pay higher prices for *organic produce*.

Permaculture

Permaculture is a system of food production which follows the patterns and features of natural ecosystems.

Permaculture practices include:

- harvesting rainwater
- crop rotation
- managing woodland.

Urban farming

Urban farming is the cultivation, processing and distribution of food in and around settlements.

The Michigan Urban Farming Initiative

- The Michigan Urban Farming Initiative in the USA aims to address problems of urban decay, poor diet and food insecurity in Detroit.
- Urban communities are encouraged to work together to turn wasteland into productive farmland, providing jobs and easier access to healthy food.

Figure 1 *Urban farming in Detroit*

Fish from sustainable sources

Almost 90% of the world's fisheries are fully or over-exploited.

Sustainable fishing involves setting catch limits (quotas) and monitoring fish breeding and fishing practices.

In Norway, salmon farms are spread out to reduce the possible spread of disease.

Meat from sustainable sources

Sustainable meat production involves small-scale livestock farms, using free-range or organic methods.

Prices may be higher in the shops but quality and animal welfare standards are higher.

Six Second Summary

- Organic farming, permaculture, urban farming and consuming fish and meat from sustainable sources can all create potential for sustainable food supplies.

 Over to you

- Make a list of the different strategies to create sustainable food supplies.
- Add **two** sentences to each strategy to explain how it creates the potential for sustainable food supplies.

Student Book
See pages 274–5

You need to know:

- how seasonal food consumption and reducing food waste and losses can create sustainable food supplies
- about Makueni in Kenya, a local scheme to increase sustainable supplies of food in an LIC or NEE.

Seasonal food consumption

In the past, food was bought from local sources when 'in season'. It is now possible in many wealthy countries to eat every type of food throughout the year.

Local food sourcing is more sustainable. It reduces both 'food miles' and our carbon footprint.

 **Big Idea**

Food miles describe the distance covered supplying food to customers.

Carbon footprint is the measurement of the greenhouse gases that each individual produces, through the direct or indirect burning of fossil fuels.

Reducing food loss and waste

Around 32% of all food produced is lost or wasted each year.

By halving the amount of food waste, the gap between food supply and demand could be reduced by 22% by 2050.

Improved food storage and distribution using refrigerated containers

Clearer food labelling, such as 'Best before' or 'Use by'

Reducing food waste

Using sealed plastic bags to make fresh food last longer

More sensible approach to using food that is past its 'Sell by' date

The Makueni Food and Water Security Programme

The programme provided direct help to two small villages and Kanyenoni Primary School in Makueni County, Kenya.

The programme included:

- improving water supply by building sand dams for each village
- providing a reliable source of water for crops and livestock
- a training programme to support local farmers
- growing trees to reduce soil erosion.

Sand dams store water in the ground, filtering and cleaning the rainwater as it soaks into the soil. They are cost-effective and sustainable.

The project has been very successful.

- Crop yields and food security have increased.
- Water-borne diseases have been reduced.
- Less time is wasted fetching water.

 Six Second Summary

- Seasonal food consumption and reducing food waste can make food supplies more sustainable.
- Sand dams have helped to improve food security in Makueni County, Kenya.

 Over to you

Create a mind map to show how the Makueni Food and Water Security Programme has increased sustainable supplies of food in that area.

Global water supply

Student Book
See pages 276–7

You need to know:

- how water surplus and deficit vary around the world
- why there is increasing global water consumption
- how different factors affect water availability.

Water surplus and deficit

Regions with a **water surplus** have a supply of water which exceeds demand.

Other regions have a **water deficit**, where demand exceeds supply.

Areas of water deficit may have:

- low rainfall
- high densities of population and/or industry, increasing demand.

Key

Water deficit

Water surplus

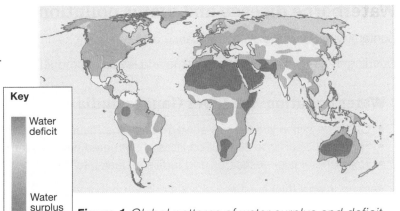

Figure 1 *Global patterns of water surplus and deficit*

Water security/insecurity

Water security means having access to enough clean water to sustain well-being, good health and economic development.

Water insecurity describes the situation where regions do not have access to sufficient water supplies.

Countries begin to experience **water stress** when less than 1700 m³ is available per person per year. Regions with high water stress include several Caribbean islands and the Middle East.

Why is water consumption increasing?

- The growth of the world's population means more water is needed.
- Economic development results in higher consumption.
- Changes in lifestyle and eating habits, e.g. more irrigation to produce food.
- Water is required for increasing energy production.

Factors affecting water availability

- *Geology* – infiltration of water through permeable rock builds up groundwater supplies.
- *Climate* – regions with high rainfall usually have surplus water.
- **Over-abstraction** – pumping water out of the ground faster than it is replaced by rainfall causes wells to dry up and lowers water tables.

- *Pollution* – increasing amounts of waste and growing use of chemicals in farming have led to higher levels of pollution.
- *Limited infrastructure* – poorer countries may lack pumping stations and pipes for transporting water to areas of need.
- *Poverty* – many poorer countries lack mains water or only have access to shared water supplies.

Six Second Summary

- Patterns of water surplus and water deficit vary around the world.
- Reasons for increasing water consumption include economic development and rising populations.
- Climate, geology and over-abstraction are some factors which affect water availability.

Over to you

Make a list of: **one** place with water surplus, **two** reasons for increasing water consumption and **three** factors affecting water availability. Add a sentence to explain each one.

You need to know:

- the impacts of water insecurity on pollution levels, food production, industry and conflict.

Student Book
See pages 278–9

Waterborne disease and water pollution

Contaminated drinking water can cause diseases such as cholera.

Queuing to get clean water wastes time and levels of productivity, and reduces time spent at school.

Water pollution: the River Ganges, India

- Over one billion litres of raw sewage enter the River Ganges each day.
- Factories discharge 260 million litres of untreated wastewater into the river daily.
- Toxic chemicals, pesticides and fertilisers leak into the river.

Bathing in and drinking the river's water have become very dangerous.

Food production

Agriculture uses 70% of global water supply and suffers the most from water insecurity.

The River Nile is Egypt's primary source of water. Climate change and the demands of countries upstream are expected to reduce its flow by 90% by 2100. Egypt currently has to import 60% of its food.

Industrial output

Growth of manufacturing industry, particularly in NEEs, is making increasing demands on water supplies.

Water shortages cost China US$40 billion in lower industrial production.

Water conflict

Water sources, such as rivers and groundwater aquifers, cross national and political borders. Issues such as reservoir construction and pollution can impact on more than one country and create conflict.

Turkey built a large number of dams on the Tigris and Euphrates Rivers, causing anger in Iraq and Syria.

Lake Chad has shrunk to 5% of its former size, due to climate change and over-abstraction.

The Nile flows through eight countries. Egypt will not allow the other seven countries to affect the Nile's flow (for example, build dams). This causes great tension in the region as countries argue over water rights.

Figure 1 *Some of the world's potential water conflict zones*

Six Second Summary

Impacts on water insecurity include: waterborne disease, water pollution, food production, industrial output and conflict.

Over to you

Create a Venn diagram of social, economic and environmental problems caused by water insecurity.

How can water supply be increased?

Student Book
See pages 280–1

You need to know:

- what strategies are used to increase water supply
- how each strategy works to increase water supply.

Diverting water and increasing storage

Water supplies can be artificially diverted and stored for use over longer periods. In Oklahoma, USA, rainfall is infrequent but heavy. Surface water quickly evaporates. So it is collected and diverted, and stored in underlying alluvial soils.

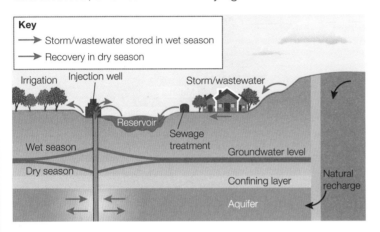

Figure 1 *Aquifer storage and recovery*

Water transfer

Schemes move water from areas of surplus to areas of deficit.

China is building three canal systems to transfer water from the Yangtze River in the south to the Yellow River Basin in the arid north. The controversial western route involves building dams and tunnels through the Bayankela Mountains.

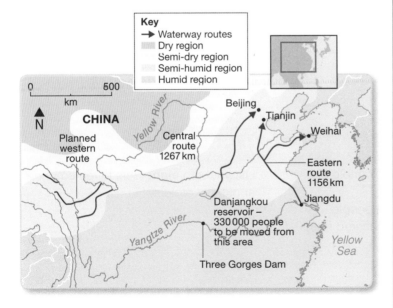

Figure 2 *China's south–north water transfer scheme*

Dams and reservoirs

Dams bring advantages:
- They control river flow by storing water in reservoirs.
- The control of water flow enables it to be transported and used for irrigation.
- They help to prevent flooding.

However, large dams:
- are expensive
- can lead to the displacement of large numbers of people
- may reduce the flow of water downstream.

In hot and arid regions, reservoirs can lose a lot of water through evaporation.

Desalination

- Desalination means removing salt from seawater to produce fresh water.
- This is a very expensive process.
- It is used only when there is a serious shortage of water with few alternatives to increase water supply.
- Both Saudi Arabia and UAE have developed desalination plants.

Six Second Summary

- Strategies used to increase water supply include: diverting supplies and increasing storage, dams and reservoirs, water transfers and desalination

Over to you

Design a poster to show how water supply can be increased.

The Lesotho Highland Water Project

Example

Student Book
See pages 282–3

You need to know:

- about the Lesotho Highland Water Project, a large-scale water transfer scheme
- the advantages and disadvantages of this development.

Lesotho is a highland country, surrounded by South Africa. It is heavily dependent economically on South Africa.

Despite food insecurity, Lesotho has a water surplus mainly due to high rainfall in its mountainous regions.

What is the Lesotho Highland Water Project?

- It is a huge water transfer scheme aimed to help solve the water shortage in South Africa.
- 40% of the water from the Segu (Orange) River in Lesotho will eventually be transferred to the River Vaal in South Africa.
- It involves the construction of dams, reservoirs and pipelines as well as roads, bridges and other infrastructure.

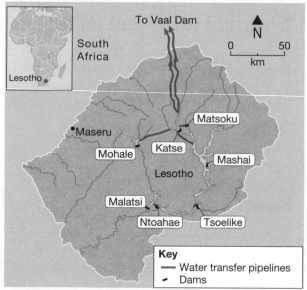

Figure 1 *Map of the Lesotho Highland Water Project*

What are the advantages and disadvantages of the scheme?

Advantages for Lesotho

- Provides 75% of Lesotho's GDP.
- Supplies all Lesotho's hydro-electric (HEP) requirements.
- Sanitation coverage will increase from 15 to 20%.

Disadvantages for Lesotho

- Building the first two dams displaced 30 000 people.
- Destruction of a unique wetland ecosystem.
- Corruption has prevented money reaching those affected by the construction.

Advantages for South Africa

- Provides water to an area with regular droughts.
- Fresh water reduces the acidity of the Vaal River Reservoir.
- Provides safe water to an extra 10% of the population.

Disadvantages for South Africa

- Costs are likely to reach US$4 billion.
- 40% of water is lost through leakages.
- Corruption has plagued the whole project.

 **Six Second Summary**

- The Lesotho Highland Water Project is an example of a large-scale water transfer scheme.
- It generates energy and water supplies but people have been displaced and the project has suffered from corruption.

Over to you

List your top **eight** facts to remember about this project.

Sustainable water supplies

Student Book
See pages 284–5

You need to know:

- how different strategies can help to create sustainable water supplies.

What is sustainable water supply?

Sustainable approaches to water supply focus on:

- management of water resources
- reducing waste and excessive demand.

Groundwater management

Groundwater is stored in underground aquifers.

To ensure sustainability, water abstraction (loss) must be balanced by recharge (gain).

If groundwater levels fall, water can become contaminated, making expensive water treatment necessary.

> **Participatory Groundwater Management (PGM), India**
>
> The PGM scheme involves:
> - training local people to monitor rainfall and groundwater levels
> - helping farmers to plan how much water to use for irrigation
> - encouraging farmers to plant crops to fit in with periods when water is available.
>
> Through PGM, rural communities have balanced water supply and demand using sustainable practices.

Using grey water

- **Grey water** is taken from bathrooms and washing machines.
- If used within 24 hours it contains fertiliser for plants.
- Water from toilets cannot be used in this way.
- In Jordan, 70% of the water used for irrigation and gardens is grey water.

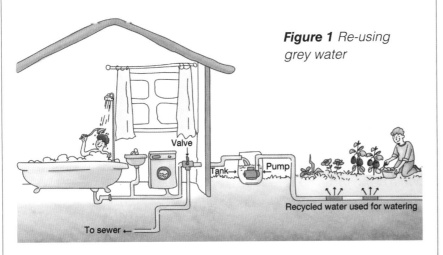

Figure 1 *Re-using grey water*

Recycling

Water recycling involves re-using treated wastewater for purposes like irrigation and industry.

- In Kolkata, India, sewage water is re-used for fish farming and agriculture.
- Some nuclear power plants – such as in Arizona, USA – use recycled water for cooling.

Ways to conserve water

- Reduce leakages
- Improve public awareness of the importance of saving water
- Water meters
- Prevent pollution
- Turn off tap when brushing teeth

Six Second Summary

Water conservation, groundwater management, recycling and using grey water can all help to create sustainable water supplies

Over to you

Write the water supply strategies on brightly coloured pieces of card or on sticky notes. Place them around your home to help remind you of these strategies.

The Wakel River Basin project

Student Book
See pages 286–7

You need to know:

- about the Wakel River Basin project, an example of a local scheme in an LIC or NEE
- how the Wakel River Basin project increases sustainable supplies of water.

Example

The Wakel River Basin is in north-west India in the south of Rajasthan – the driest and poorest part of India. The rainfall of less than 250 mm per year quickly soaks away or evaporates.

What are the issues with water supply?

- Water management in the region has been poor.
- Over-use of water for irrigation has led to waterlogging and salinisation.
- Over-abstraction from unregulated pumps has resulted in falling water tables in aquifers.
- Some wells have dried up.

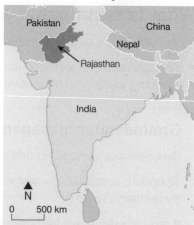

Figure 1 *Location of Rajasthan*

Increasing water supply in the Wakel River Basin

The United States Agency for International Development has been working with local people in the Wakel River Basin. The project aims to improve water security and overcome the problems of water shortages by encouraging greater use of rainwater harvesting techniques.

- *Taankas* – underground storage systems which collect surface water from roofs.
- *Johed* – small earth dams that capture rainwater.
- *Pats* – irrigation channels that transfer water to the fields (Figure **2**).

How does the *pat* system work?

A small dam called a bund diverts water from the stream towards the fields. Villagers take turns to irrigate their fields in this way. Maintenance is done by the villager whose turn it is to receive the water.

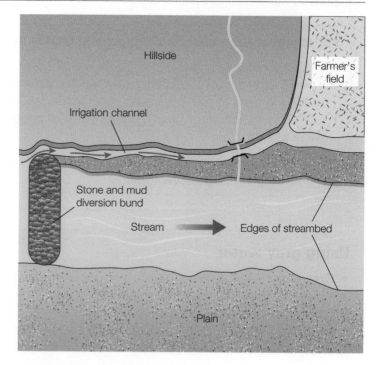

Figure 2 *The* pat *irrigation system*

Increasing public awareness

Education is used to increase awareness of the need for communities to work together to conserve water.

 Six Second Summary

- Earth dams, irrigation channels and public awareness have helped water security to increase in the Wakel River Basin.

 Over to you

- Write down **three** ways in which the Wakel River Basin project is increasing sustainable supplies of water.
- For each way, add the words 'this means that …' and finish the sentence to explain why it is increasing sustainable supplies of water.
- Keep adding 'this means that' as long as you can to add depth to your answer.

Student Book
See pages 288–9

You need to know:

- how energy consumption and supply vary around the world
- why there is increasing energy consumption
- how different factors affect energy supply.

Global energy consumption and supply

A country's **energy security** depends on its supply and consumption. If supply exceeds demand, it has an *energy surplus*. If demand exceeds production, it has an *energy deficit* and the country is *energy insecure*.

- Consumption is highest in North America and parts of the Middle East.
- Consumption is lowest across most of Africa and parts of south-east Asia.
- North America has large coal reserves.
- Russia has large reserves of natural gas and oil.
- Sub-Saharan Africa depends on overseas TNCs to exploit reserves.

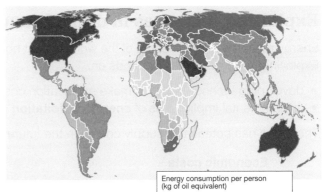

Figure 1 *Global energy consumption*

Energy consumption per person (kg of oil equivalent)
- More than 10 000
- 5001–10 000
- 2501–5000
- 1001–2500
- 501–1000
- 0–500

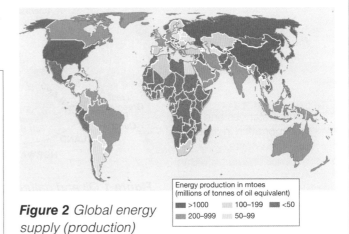

Figure 2 *Global energy supply (production)*

Energy production in mtoes (millions of tonnes of oil equivalent)
- >1000
- 200–999
- 100–199
- 50–99
- <50

What factors affect supply?

Technology

- Technology makes energy sources in difficult environments exploitable.

Political factors

- Political instability in the Middle East means countries seek alternative energy sources.
- The UK government has cut subsidies for renewable energy.

Costs of exploitation and production

- Some energy sources are costly to exploit.
- Nuclear power stations are expensive to build.

Physical factors

- Geology determines the availability of fossil fuels.
- **Geothermal energy** is produced in areas of tectonic activity.

Climate

- The amount of sun and wind influence the availability of solar energy and wind energy.
- HEP needs a suitable dam site in mountainous areas with high rainfall.

Why is consumption increasing?

- As countries develop economically, their demand for energy supplies rises.
- As global population increases, so does energy consumption.
- Technology helps produce cheaper energy.

Six Second Summary

- Energy consumption and supply vary globally.
- Physical and political factors affect energy supply.
- Increasing energy consumption results from development and rising populations.

Over to you

Use the maps to describe the global distribution of energy consumption and supply.

Student Book
See pages 290–1

You need to know:

- the impacts of energy insecurity.

Exploiting resources in difficult and sensitive areas

Energy resources exist in some of the world's most hostile and environmentally sensitive regions. Exploiting these resources depends on:

- developing technologies that make exploitation cost-effective
- environmental implications of **energy exploitation** in areas that could easily be damaged.

The Arctic has potential to supply energy in the future, but there are costs:

Economic costs

Exploitation is difficult and expensive.

People demand higher wages to work there.

Long distances and limited transportation increase transport costs.

Environmental costs

The environmental consequences of an oil spill would be catastrophic for the fragile Arctic ecosystem.

Strict environmental controls are needed.

Drilling equipment may sink during the summer thaw.

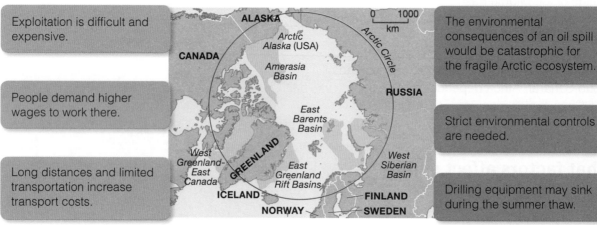

Figure 1 *Oil and natural gas resources in the Arctic (yellow shading)*

Impacts on food production

- Using biofuels like maize and sugar cane for energy have led to increased food prices.
- Biofuels are often grown on land previously used for growing food crops.
- In some LICs, collecting wood for fuel reduces time spent on food production.

Impacts on industry

- Some countries suffer from shortfalls in electricity production.
- Energy shortages have led to the closure of more than 500 companies in the industrial city of Faisalabad, Pakistan.

Potential for conflict

- Shortages of energy can lead to political conflict.
- The Middle East produces 40% of the world's gas and 56% of its oil.
- The Gulf and Iraq wars in the 1990s and 2000s were influenced by the West's fear of a global oil shortage and rising prices.

Six Second Summary

Impacts of energy insecurity include:
- exploration of difficult and environmentally sensitive areas
- economic and environmental costs
- food production   • industrial output   • conflict

Over to you

Look at the bullet points in the Six Second Summary.

Write one sentence to explain each one.

Student Book
See pages 292–3

You need to know:

- what strategies are used to increase energy supply
- how each strategy works to increase energy supply.

What are the options for increasing energy supplies?

Every day we're using more and more energy!
There are two main options for increasing future energy supplies:

- increase the use of renewable energy sources
- exploit non-renewable fossil fuels and develop the use of nuclear power.

To avoid relying too much on one source, most countries use several energy types –
an **energy mix**. Most countries use a mix of renewable and non-renewable sources.

Renewable energy sources

Renewable energy source	Can it increase energy supplies?
Biomass	Using land to grow biofuels rather than food crops is controversial. Fuelwood supplies are limited.
Wind	Unpopular with some, but considerable potential.
Hydro (HEP)	An important energy source in several countries. It currently contributes 85% of global renewable electricity.
Tidal	There are few tidal barrages due to high costs and environmental concerns.
Geothermal	Limited to tectonically active countries.
Wave	There are many experimental wave farms but costs are high and there are environmental concerns.
Solar	Energy production is seasonal. Great potential in some LICs which tend to have lots of sunshine.

Non-renewable energy sources

Fossil fuels

Although stocks of fossil fuels are limited, there are still plenty left. But, at some point, they will become scarce and their price too high. Eventually, they will run out.

- Fossil fuels include coal, gas and oil.
- They remain important for electricity production.
- Carbon capture can help overcome environmental impacts.

Nuclear power

- Radioactive waste can remain dangerous for over 1000 years.
- Despite a good safety record, there is considerable opposition, and it's an expensive choice.
- Low cost of uranium as little is used.

Six Second Summary

Strategies to increase energy supply include renewable and non-renewable sources of energy.

Over to you

Revisit this page in two days' time and remind yourself of anything you have forgotten.

Gas – a non-renewable resource

Student Book
See pages 294–5

Student Book
See pages 294–5

You need to know:

- that natural gas is an **example** of a fossil fuel
- the advantages and disadvantages of extracting natural gas.

Example

How is natural gas formed?

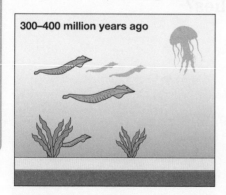

300–400 million years ago

Remains of tiny sea plants and animals buried on the ocean floor by sand and sediment.

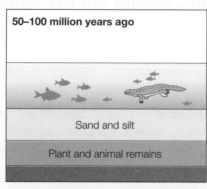

50–100 million years ago

Sand and silt

Plant and animal remains

Over millions of years, the remains were buried deeper. Enormous pressure and heat converted organic material into hydrocarbons (oil and gas).

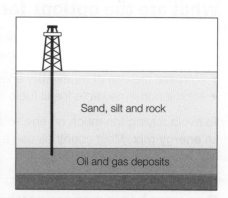

Sand, silt and rock

Oil and gas deposits

The natural gas seeped through cracks in overlying rocks and collected in reservoirs. It is from these reservoirs that natural gas is extracted.

Figure 1 *The formation of oil and natural gas*

Where is natural gas found?

Nearly 60% of known natural gas reserves are in Russia, Iran and Qatar, with reserves expected to last 54 years at the current rate of production. Recent technology has allowed **shale gas** to be extracted by a controversial process called *fracking*.

Extracting natural gas

Advantages

- Emits 45% fewer CO_2 emissions than other fossil fuels.
- Lower risk of environmental accidents than oil.
- Can be transported easily via pipelines or by tankers.

Disadvantages

- Some gas reserves are in politically unstable countries.
- Wastewater and chemicals from fracking can contaminate groundwater.
- Contributes to global warming by producing CO_2 and methane emissions.

Extracting natural gas in the Amazon

The Camisea project began in 2004 to exploit a huge gas field in the Amazonian region of Peru.

Advantages

- Peru could make several billion dollars in gas exports.
- It provides employment opportunities.
- It could save Peru up to US$4 billion in energy costs.

Disadvantages

- Deforestation will affect the Amazon.
- The project could affect traditional lifestyles of indigenous tribes.
- Local people have no immunity to disease introduced by developers.

Six Second Summary

- Advantages of extracting natural gas include employment and lower risk of environmental accidents.
- Disadvantages include deforestation and contribution to global warming.

Over to you

Create **ten** questions about everything you have learnt about energy. Test yourself in a few days.

Sustainable energy use

Student Book
See pages 296–7

You need to know:

- how different strategies can conserve energy and work towards a sustainable resource future.

What is a sustainable energy supply?

A sustainable energy supply involves balancing supply and demand. It requires individual actions and decisions made by businesses and governments.

Energy demand can be reduced by increasing **energy conservation**.

Reducing the use of fossil fuels will help reduce our *carbon footprint*.

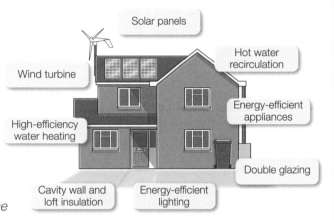

Figure 1 *Energy conservation in the home*

Sustainable energy developments in Malmo, Sweden

Malmo's Western Harbour is one of the world's best examples of sustainable urban redevelopment.

Homes and workplaces

- All 1000 buildings in the district use 100% renewable energy.
- Solar tubes on the outside of buildings produce hot water which can be stored in aquifers.
- Energy comes from photovoltaic panels on roofs, a 2MW wind turbine, and biogas from local sewage and rubbish.

Transport

- From 2019, all buses will run on a mixture of biogas and natural gas.
- Cyclists have priority at crossroads.
- Buses and water taxis provide frequent public transport.

Reducing energy demand

Ways of reducing energy demand can include:

- financial incentives
- raising awareness of the need to use energy efficiently
- greater use of off-peak energy
- using less hot water for domestic appliances.

How can technology increase efficiency of fossil fuels?

Vehicle manufacturers are using technology to design more fuel-efficient cars. These developments include improved engines and aerodynamic designs.

Development of electric and hybrid cars – in the USA, electric cars could reduce the use of oil for transport by up to 95%.

Development of biofuel technology – Brazil has reduced its petrol consumption by 40% since 1993 by using a *biofuel* called *sugar cane ethanol*. However, growing biofuels rather than food crops is a controversial issue.

 Six Second Summary

Energy conservation can occur by designing homes, workplaces and transport for sustainability, reducing demand, and through the use of technology.

Over to you

Write down **three** ways that energy use could be made more sustainable. For each one, write a reason why that method will make energy use more sustainable.

Student Book
See pages 298–9

You need to know:

- about the Chambamontera micro-hydro scheme – an example of a local renewable energy scheme in an LIC or NEE
- how it provides sustainable supplies of energy.

Example

Why does Chambamontera need a sustainable energy scheme?

Chambamontera is an isolated community in the Andes Mountains of Peru.

- Nearly half the population survive on just US$2 a day.
- Steep slopes and rough roads make Chambamontera very isolated.
- Due to the low population density, it was uneconomic to build an electricity grid to serve the area.

Figure 1 *Location of Chambamontera*

What is the Chambamontera micro-hydro scheme?

The high rainfall, steep slopes and fast-flowing rivers make this area ideal for exploiting water power as a renewable source of energy.

The total cost of the scheme was US$51 000. There was some government money invested from Japan, but the community had to pay part of the cost. Credit facilities were made available to pay for this.

How has the local community benefited?

- Provides renewable energy
- Has low maintenance and running costs
- Has little environmental impact
- Used local labour and materials.

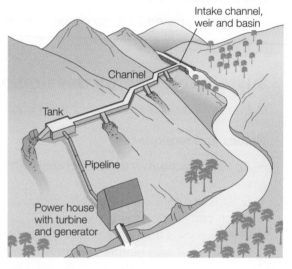

Figure 2 *How the scheme works*

Reduced rural–urban migration, so the population has grown.

Business development is possible, as piped water can drive small machines for coffee de-husking and processing.

Regulating the flow of water has reduced the danger of flooding.

Benefits to the local community

Reliable electricity for refrigeration, light and other uses like computers and entertainment.

Improved school facilities and the possibility of doing schoolwork at home after dark.

Less need to burn wood as a source of heat, so reduced deforestation and risk of soil erosion.

Six Second Summary

- The Chambamontera micro-hydro scheme exploits water power as a renewable source of energy.
- It brings benefits such as electricity and business development.

Over to you

Think of a rhyme or a mnemonic to help you to remember the name or the Chambamontera micro-hydro scheme.

Unit 3: Geographical applications

Issue evaluation and Fieldwork

Paper 3 is called 'Geographical applications'. Instead of revising and learning by heart, it is much more practical and you must use your knowledge, understanding and skills. It has two sections:

Your exam

Section A Issue evaluation and Section B Fieldwork make up Paper 3: Geographical applications.

Paper 3 is a one-and-a-quarter hour written exam and makes up 30 per cent of your GCSE. The whole paper carries 76 marks (including 6 marks for SPaG) – questions on Section A will carry 37 marks; questions on Section B will carry 39 marks.

In your final exam you will have to answer all questions.

Section A: Issue evaluation

This section involves learning about a new issue which you won't have studied before. It'll be contained in a resource booklet which you'll be given about 12 weeks before the exam.

- The **resource booklet** will consist of maps, diagrams, graphs, statistics, photographs, sketches, text, and quotes from different groups about an issue.

- The **exam** will assess your ability to make sense of the resource booklet. In those 12 weeks, you'll be able to read, understand, and prepare for the exam.

The issue could be about anything, but will be linked to the compulsory sections of subject content. These are:

- **Unit 1:** Tectonic hazards, Weather hazards (see Figure **1**), Climate change, Ecosystems, and Tropical rainforests. An example of an issue based on tectonic hazards appears in 23.2.

- **Unit 2:** Urban issues and challenges and The changing economic world.

Figure 1 *A tropical storm in Puerto Rico in Autumn 2017. How should Puerto Rico try to recover from damage done? This could be the topic of an Issue evaluation.*

Section B: Fieldwork

You must revise your two fieldwork topics. It will help if you practise answering exam questions on fieldwork so that you're used to applying what you learnt on your two topics.

Your two fieldwork topics will be:

- a physical geography topic, where you carried out an investigation into rivers, coasts or glaciated landscapes and processes

- a human geography topic, where you've investigated a rural or an urban place (like the one in Figure **2**).

The questions will be about two types of fieldwork:

- fieldwork you have carried out yourself – i.e. *familiar* fieldwork

- fieldwork where you apply what you learned to new places – i.e. *unfamiliar* fieldwork.

Figure 2 *A housing estate in east London. This could be the topic of an fieldwork investigation.*

You need to know:

- how to prepare for the exam in Paper 3 (the Issue evaluation and Fieldwork).

What makes Unit 3 different?

Chapters 1 to 22 in this book help you to prepare for Paper 1 (Living with the physical environment) and Paper 2 (Challenges in the human environment). You prepare by revising topics you've done in class and at home.

Unit 3 is different. The exam is less about learning, and more about preparation. Half of it is an *Issue evaluation* – a topic about which different people have different views, which you have to evaluate (or weigh up).

- It's based on a resource booklet which you'll receive 12 weeks before the exam.
- In those 12 weeks, you'll be able to read, understand, weigh up different views, and prepare for the exam.

The other half is based on the two days of *fieldwork* that you've carried out. You can prepare by reading through your fieldwork, and by reading Unit 24 in this book.

The Exam

The exam lasts for 1 hour 15 minutes, which is split between the Issue evaluation and Fieldwork sections.

It has 76 marks in total.

- **Section A** The Issue evaluation, has 37 marks, 3 of which are for spelling, punctuation and grammar.
- **Section B** The Fieldwork section, has 39 marks, 3 of which are for spelling, punctuation and grammar.

So, aim for a mark a minute!

That sounds like a pressurised exam – but remember you'll know the content of the resource booklet already. You'll also probably know which project or option you'll choose for the last question, so you can prepare.

The Issue evaluation

The Issue evaluation is based on a six-page resource booklet. You'll receive this 12 weeks before the exam.

- It will contain geographical information that you probably won't have seen before about an issue somewhere in the world.
- It's likely that you won't know much about this place. Don't worry – it isn't your knowledge of the place that's being assessed, but your ability to understand the issue.
- To help you understand the issue, the resource booklet contains text, maps, photos, diagrams and data. The exam will assess your ability to study these and make sense of them.
- Towards the end of the resource booklet, you'll see three or four projects or options, which will give alternative ways of dealing with the issue. The options will be real ones, not fictional.
- The final question in the exam will ask you to select one of these projects or options. There won't be a 'right' or 'wrong' answer. You'll be marked on how well you argue the reasons you give for your choice.

The resource booklet will also contain key words and concepts that you'll understand from Units 1 and 2. That's intentional – examiners will want you to make links between topics you've studied. This is called being *synoptic* – making links with what you know and understand.

> **Remember!**
>
> It's your ability to understand the issue that's important.

A sample resource booklet follows in 23.2, like the one you'll use in the exam. It's organised into questions.

- The first question introduces you to the place or issue – the example in this topic is Christchurch, New Zealand.
- The next questions will explore the issue – you'll see that Christchurch was hit by several earthquakes in 2010 and 2011, which seriously damaged some parts of the city. Christchurch now faces an issue – what kind of city should be rebuilt, if at all? The resource booklet will take you through this.

What will the Issue evaluation topics be?

The topic for the Issue evaluation will differ each year. You won't know until the resource booklet is given out. It will always be about *core topics*, not options. Core topics are:

- **Unit 1 Section A** The challenge of natural hazards (natural hazards, tectonic hazards, weather hazards, climate change)
- **Unit 1 Section B** The living world (ecosystems, tropical rainforests)
- **Unit 2 Section A** Urban issues and challenges (case study of a major city in *either* a low-income country *or* a newly emerging economy, *and* a case study of a major city in the UK)
- **Unit 2 Section B** The changing economic world (case study of *either* a low-income country *or* a newly emerging economy, *and* economic futures in the UK)
- **Unit 2 Section C** The challenge of resource management (resource management)

Christchurch CBD after the earthquakes of 2010–11. People are questioning whether the city is worth rebuilding at all.

The issue

- Christchurch, New Zealand, was affected by a series of earthquakes between 2010 and 2011, with smaller ones in 2016.
- Much of the city centre and suburbs east of the city were destroyed in 2011.
- There is debate about how Christchurch should be rebuilt. Who has the best ideas about Christchurch's future?

Figure 1 The earthquakes in Christchurch

Figure 1a Factfile on Christchurch, New Zealand

Fact file

- Christchurch is located on South Island, New Zealand.
- It is New Zealand's third largest city, and the largest city on South Island.
- It is surrounded by a region called Canterbury, which is mainly farmland and small towns (with a lower population density).
- About 375 000 people live in Christchurch itself, and another 200 000 in the Canterbury region.

Figure 1b The earthquakes of 2010–11

- Several minor earthquakes occur every day in New Zealand.
- There were three big earthquakes between September 2010 and June 2011, and others that were smaller (see **Figure 1c**).
- Earthquakes that occur in large numbers like this are known as 'earthquake swarms'.

Figure 1c The area around Christchurch

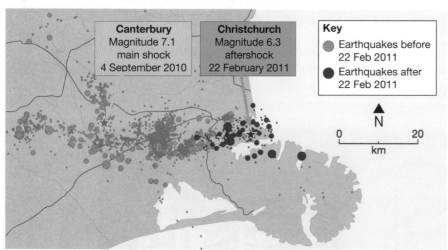

Canterbury	Christchurch	Key
Magnitude 7.1 main shock 4 September 2010	Magnitude 6.3 aftershock 22 February 2011	○ Earthquakes before 22 Feb 2011 ● Earthquakes after 22 Feb 2011

0   N   20
km

The Christchurch region showing the swarm of earthquakes that occurred between September 2010 and June 2011

Time sequence of the main earthquakes

- First earthquake – September 2010 in Canterbury (7.1 on the Richter scale).
- Second earthquake – February 2011 in Christchurch (6.3) which affected the city itself.
- Third earthquake – June 2011 in Christchurch (also 6.3). Many people and businesses had been evacuated by this time.
- A larger earthquake (7.8) occurred in 2016, further north in the Marlborough wine-producing region – this only affected Christchurch slightly. However, areas around Christchurch were affected. See **Figure 3f**.

Figure 2 What caused the earthquakes?

New Zealand lies across a plate margin. The plates are active and cause thousands of earthquakes every year.

Figure 2a

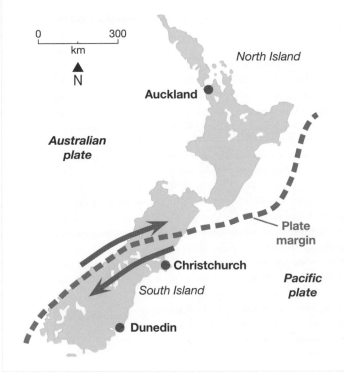

Figure 2b Earthquake prediction and protection

- A lot is known about where earthquakes are likely to happen.
- There is no known way of predicting when they will happen.
- People can prepare for earthquakes in different ways.
- Buildings can be designed to withstand earthquakes in different ways.

The plate margin which caused the earthquakes

Figure 3 Impacts of the earthquakes in Christchurch

Figure 3a

The city centre is the oldest part of Christchurch. Two kinds of buildings suffered most in the earthquakes of 2011: the oldest buildings (e.g. Christchurch Cathedral, (**Figure 3b**) and the tallest buildings – out of 220 buildings over five storeys high, half have had to be demolished.

Further east, many houses were destroyed. They had been built on softer sands (which makes poor foundations) near the River Avon (**Figure 3c**).

The 2010–11 earthquakes were New Zealand's most expensive natural disaster costing NZ$20 billion in total. New Zealand's total GDP was NZ$200 billion in 2010.

Figure 3b

Christchurch Cathedral, one of the city's oldest buildings, was badly damaged in the earthquake in February 2011. In 2017, it was still derelict.

Figure 3c

Housing in Avonside, east of Christchurch city centre

Figure 3d

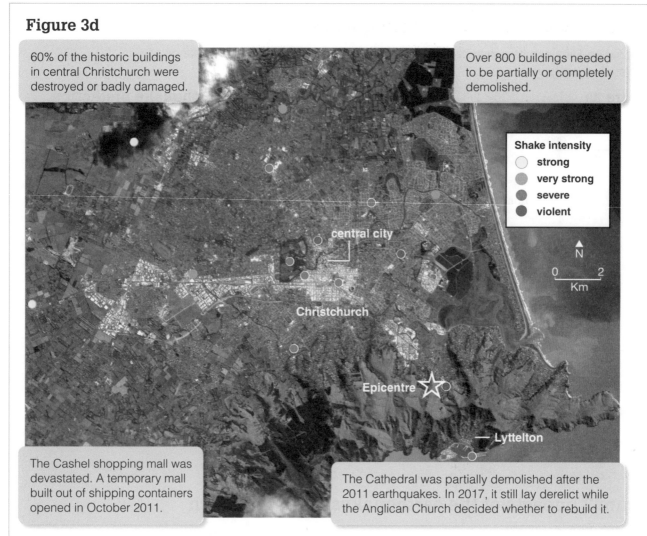

60% of the historic buildings in central Christchurch were destroyed or badly damaged.

Over 800 buildings needed to be partially or completely demolished.

Shake intensity
- strong
- very strong
- severe
- violent

The Cashel shopping mall was devastated. A temporary mall built out of shipping containers opened in October 2011.

The Cathedral was partially demolished after the 2011 earthquakes. In 2017, it still lay derelict while the Anglican Church decided whether to rebuild it.

Areas of Christchurch worst affected by the second earthquake on 21 February 2011

Figure 3e

	3 September 2010	**21 February 2011**	**13 June 2011**
Location	Canterbury	Christchurch urban area	East of Christchurch
Richter Scale	7.1	6.3	6.3
Deaths	0	181 (115 of these were in the Canterbury TV building)	0
Injured	100	6000–7000	46
Cost of damage	NZ$3 billion	NZ$15 billion	NZ$60 million
Buildings	Many buildings were weakened, but only a few were destroyed in the city centre.	Caused major damage. 1000 buildings in the city centre and to the east were destroyed or were demolished later.	Many buildings in the city centre were already damaged or had been evacuated and demolished.
Other points	Affected Canterbury; some damage in Christchurch.	Affected the city centre badly.	

Comparing the effects of the three largest earthquakes 2010–11

Figure 4 Should Christchurch be abandoned?

Figure 4a Future earthquakes

It is very likely that Christchurch will be affected by more earthquakes in future. Its location is in an area of active movement. Already 70 000 people – 20% of Christchurch's population – have left the city temporarily since the 2010–11 earthquakes while rebuilding takes place.

- There have been over 100 000 aftershocks since September 2010.
- Scientists predicted there was a 72% chance that Christchurch would be struck by an earthquake with a magnitude between 5 and 5.4 between 2012 and 2013.

Figure 4b Does Christchurch have a future?

An article suggesting that Christchurch should be abandoned

Parker dismisses abandoning city

Christchurch Mayor Bob Parker has hit out at suggestions that rebuilding earthquake-hit Christchurch should be abandoned.

In the Otago Daily Times yesterday, a Dunedin councillor said it was insane to rebuild Christchurch on the same site, and the money should be spent in developing Dunedin instead.

'Rebuilding Christchurch and hoping for no more earthquakes will doom Christchurch and the South Island to long-term loss of investment,' he said. 'It would be foolish to pour money into a city that could be hit by yet another big earthquake.'

Figure 4c Should Christchurch be rebuilt?

The government has divided Christchurch up into four zones: blue, red, orange and white (see map below). These show where rebuilding will or could take place.

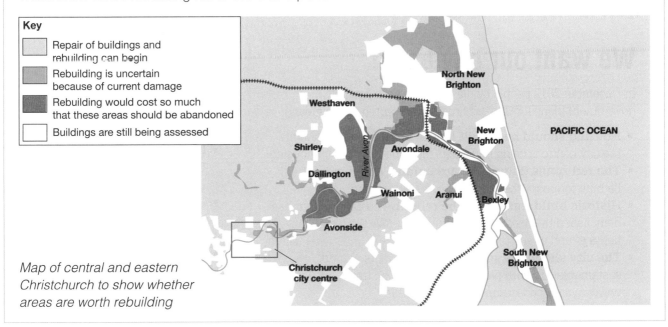

Key
- Repair of buildings and rebuilding can begin
- Rebuilding is uncertain because of current damage
- Rebuilding would cost so much that these areas should be abandoned
- Buildings are still being assessed

Map of central and eastern Christchurch to show whether areas are worth rebuilding

Figure 5 How should Christchurch be redeveloped?

Figure 5a Project 1: The Government plan

The New Zealand Government and Christchurch City Council have formed the Canterbury Earthquake Recovery Authority (CERA). It believes that Christchurch should be fully rebuilt, but with conditions:

- rebuilding offices, workplaces and shops is the priority
- many heritage buildings were too dangerous and not earthquake-proof – they should be demolished
- the earthquake is a chance to re-plan Christchurch city centre and the economy to attract high salary earners in the quaternary sector
- all buildings should be built to resist earthquakes.

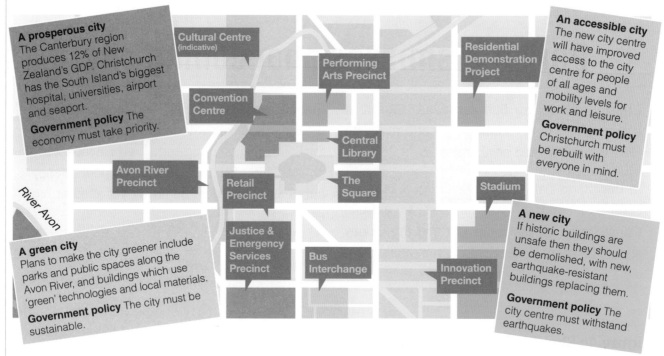

The government plan for rebuilding Christchurch city centre based on 100 000 ideas received from the public

Figure 5b Project 2: The People's plan

We want our city back!

In December 2012 people held a protest march against government plans for Christchurch. They wanted the following:

- **Housing should be rebuilt before businesses.** Although the city centre is being replanned, there are no plans for housing.
- **The red zones (Figure 4c) are unfair.** New homes should be built for people before they are forced to leave.
- **Historic buildings should be rebuilt.** The government has demolished the finest heritage buildings. 'We are losing our heritage and character' said one resident.
- **The city should be sustainable.** Residents want a sustainable city with parks, improved public transport, walkways for pedestrians and cycle paths.

Figure 5c Project 3: Abandon Christchurch and build a new city

This plan claims that no city is earthquake-proof, so rebuilding Christchurch is a waste of money.

- No new building plans should be made for Christchurch – either residential or commercial.
- Companies should be offered new offices in other cities in New Zealand.
- All residents should be compensated for their houses and asked to leave.

Exam questions

- These questions are based on 23.2 in the resource booklet (pages 166–170).
- You should take 37 minutes for this section.
- There are 37 marks, which includes 3 marks for spelling, punctuation and grammar (SPaG).
- The mark schemes are on the following pages – no peeking until you've done the questions! Use this to self-assess your answers to the exam questions below.
- When you mark your answers, read the mark scheme carefully to see how to improve.
- Then do a second attempt to see how close you can get to full marks.

Question 1 (8 marks)

1.1 Christchurch is in which **one** of the following countries?

- **A** Australia ☐
- **B** New Zealand ☐
- **C** France ☐
- **D** Nigeria ☐

[1 mark]

1.2 Christchurch is in which **one** of the following locations?

- **A** In the northern hemisphere ☐
- **B** On the Equator ☐
- **C** In the southern hemisphere ☐
- **D** In Antarctica ☐

[1 mark]

1.3 Suggest why the earthquake in February 2011 had bigger impacts than that of September 2010.

[6 marks]

Question 2 (9 marks)

2.1 The arrows on the diagram of the plate margin (**Figure 2a**) show that this is which type of plate margin?

- **A** Collision ☐
- **B** Conservative ☐
- **C** Constructive ☐
- **D** Destructive ☐

[1 mark]

2.2 Explain **one** way that this kind of plate margin can cause earthquakes.

[2 marks]

2.3 Study **Figure 3**. Assess the impacts of the earthquakes between 2010 and 2011 on Christchurch and the surrounding region.

[6 marks]

Question 3 (20 marks)

3.1 Explain one reason why earthquakes are likely to happen again in Christchurch.

[2 marks]

3.2 Discuss the arguments for and against abandoning cities that are likely to be affected by earthquakes.

[6 marks]

3.3 Three projects have been suggested about the future of Christchurch. These are outlined in **Figure 5**.

- Which of the three projects do you think should be adopted for the people, the economy, and the environment of Christchurch?
- Use evidence from this resource booklet and your own understanding to explain why you have reached this decision.

[9 marks]

[+ 3 SPaG marks]

Mark scheme

Question 1 (8 marks)

1.1 **B** New Zealand [1 mark]

1.2 **C** In the southern hemisphere [1 mark]

1.3 Use these points to guide you. Include some (but not all) of the following points:

- The second earthquake in February 2011 was less strong than the first.
- It was in the city itself with a higher density of buildings, unlike the first which was in Canterbury (which is agricultural with a lower population/population density).
- Damage in 2011 was caused by weakening of buildings, roads etc., so the second earthquake destroyed what was already damaged in 2010. Types of damage should be referred to (e.g. cost), or examples (e.g. collapse of Christchurch Cathedral).
- Because city buildings collapsed, the cost was greater than the 2010 earthquake.

You may use or refer to material researched that goes beyond the booklet.

Level	Mark	Description
3 (Detailed)	5–6	• A detailed explanation of the impacts of the different earthquakes; much evidence from the resource booklet. • Communicates ideas with clarity.
2 (Clear)	3–4	• A sound understanding of differences between the two earthquakes. • Demonstrates clear explanation of the impacts of both earthquakes; some evidence from the resource booklet.
1 (Basic)	1–2	• A limited understanding of how or why the two earthquakes differed. May simply identify reasons unrelated to the resource booklet. • Demonstrates limited understanding of the impacts of the earthquakes; little evidence from the resource booklet.
	0	No relevant content.

Question 2 (9 marks)

2.1 **B** Conservative [1 mark]

2.2 Award 1 mark for a factor and 1 mark for development of that point. Answers could include:

- Horizontal movement **(1)** so that 'sticking' can lead to build up of force. **(1)**
- Shallow/not very deep within the crust **(1)** so the force is felt more strongly. **(1)**

2.3 Use the following points to help you. Remember that if you just describe or explain, without saying which were the *biggest impacts*, you cannot reach Level 3.

Your answer should recognise some (but not all) of the following points. It will help to categorise impacts, for example, social, economic and environmental.

- Social impacts – personal injury or death of family members; trauma; loss of housing in eastern part of the city; high repair cost for many houses; need to move away from area, loss of communities.
- Economic impacts – examples of costs of damage; damage to workplaces; jobs lost; potential for the city to lose jobs if companies decide to leave.
- Environmental impacts – for example, on the heritage buildings within Christchurch; loss of character (e.g. quotes in **Figure 5b**).

You may use or refer to material researched that goes beyond the booklet.

Level	Mark	Description
3 (Detailed)	5–6	• A detailed assessment of the most important impacts, with supporting evidence from the resource booklet. • Communicates ideas with clarity.
2 (Clear)	3–4	• A sound understanding of impacts of the earthquakes. • Demonstrates clear assessment of the relative importance of different impacts. Uses some supporting evidence from the resource booklet.
1 (Basic)	1–2	• A limited understanding of different impacts. May simply describe impacts in general terms. • Demonstrates limited or no assessment of the importance of different impacts. Little evidence from the resource booklet.
	0	No relevant content.

Question 3 (20 marks)

3.1 Award **1 mark** for a factor and **1 mark** for development of that point Answers could include:

- There have been over 10 000 aftershocks since September 2010 **(1)** so there will be more! **(1)**
- Scientists predicted a 72% chance of an earthquake with a magnitude of 5–5.4 between 2012 and 2013 **(1)** – so the odds against that happening are only 28%. **(1)**
- An earthquake occurred in 2016 **(1)** which was even bigger that 2010–11. **(1)**

3.2 Use the following points to help you. Remember that if you just describe or explain (without saying how important some reasons are for either side) you cannot reach Level 3.

Your answer should recognise some (but not all) of the following points.

- Reasons for not abandoning – loss of communities and connections to places, jobs may be lost or careers disrupted, compensation costs high, cost of levelling buildings and land.
- Reasons for not abandoning (or for rebuilding) – do not just list the opposites of the above points. Rebuilding provides jobs and economic growth; buildings may be better than those destroyed; a chance to redesign the city (using examples from **Figures 4** and **5**).

You may use or refer to material researched that goes beyond the booklet.

Level	Mark	Description
3 (Detailed)	5–6	• A detailed discussion of the reasons on both sides, with supporting evidence from the resource booklet. • Communicates ideas with clarity.
2 (Clear)	3–4	• A sound explanation of both sides of the argument. • Demonstrates clear discussion using some supporting evidence from the resource booklet.
1 (Basic)	1–2	• A limited discussion of different arguments. May be one-sided. • Demonstrates limited or no explanation. Little evidence from the resource booklet.
	0	No relevant content.

3.3 Your answer should include material from your course as well as from the *resource booklet*. You should:

- give at least three detailed reasons to reach the top of level 3, quoting evidence from the resource booklet
- refer to each of people, the economy and the environment to reach Level 3
- refer to all the options, about what is most appealing about one and why you rejected the other two
- include detailed evidence from the resource booklet to support your case
- include any evidence from the course to support your case.

You may use or refer to material researched that goes beyond the booklet.

Level	Mark	Description
3 (Detailed)	7–9	• A thorough evaluation of the effectiveness of the chosen project in terms of its benefits. • Uses a wide range of evidence to support the decision, using detailed content from different areas of the course.
2 (Clear)	4–6	• A reasonable evaluation of the effectiveness of the chosen project in terms of its benefits. • Uses some evidence from the resource booklet to support the decision, and content from different areas of the course.
1 (Basic)	1–3	• A basic evaluation of the effectiveness of the chosen project in terms of its benefits. • Uses limited evidence from the resource booklet to support the decision, using basic content from different areas of the course.
	0	No relevant content.

Spelling, punctuation and grammar [3 marks]

- **3 marks** if you spell and punctuate accurately, use rules of grammar with effective control of meaning, and use a wide range of specialist terms.
- **2 marks** if you generally spell and punctuate accurately, use rules of grammar with general control of meaning, and use a good range of specialist terms.
- **1 mark** if you spell and punctuate reasonably accurately, use rules of grammar with some control of meaning, and any errors you make do not significantly hinder meaning, and use a limited range of specialist terms.
- **0 marks** if you write nothing, or do not relate to the question, with a basic grasp of spelling, punctuation and grammar which prevents any meaning being clear.

You need to know:

- how to prepare for the Fieldwork section of the exam in Paper 3.

*Student Book
See pages 306
and 316*

What makes Unit 3 different?

The Unit 3 exam is less about revision, and more about preparation. Section A is an *Issue evaluation* (see chapter 23). Section B is based on two days' *fieldwork* that you've carried out. Prepare by reading through your fieldwork, and using this chapter.

The exam

- The exam lasts 1 hour 15 minutes. It is split between the Issue evaluation in Section A, and Fieldwork in Section B. It has 76 marks in total.
- Section B Fieldwork carries 39 marks, 3 of which are for Spelling, Punctuation and Grammar.
- You'll have 38 minutes to do this section! It's pressurised but you'll know your fieldwork already, and will have thought about how to prepare for the exam.
- See 24.2 and 24.3 for more details about the questions you could be asked in the exam.

Preparing for the Section B Fieldwork exam

You should have undertaken two days' fieldwork (one day physical, one day human), during which you should have collected primary data (data you collect yourself). The topics should be linked to the content in Units 1 and 2.

Fieldwork for AQA's GCSE Geography specification has six stages. Prepare yourself for each stage, as there could be questions about any stage. The stages are summarised in Figure **1**. Also see 24.4–24.9.

Before you begin revision of your fieldwork, make sure you know about the fieldwork you have carried out and how it will be assessed. Complete the checklist below.

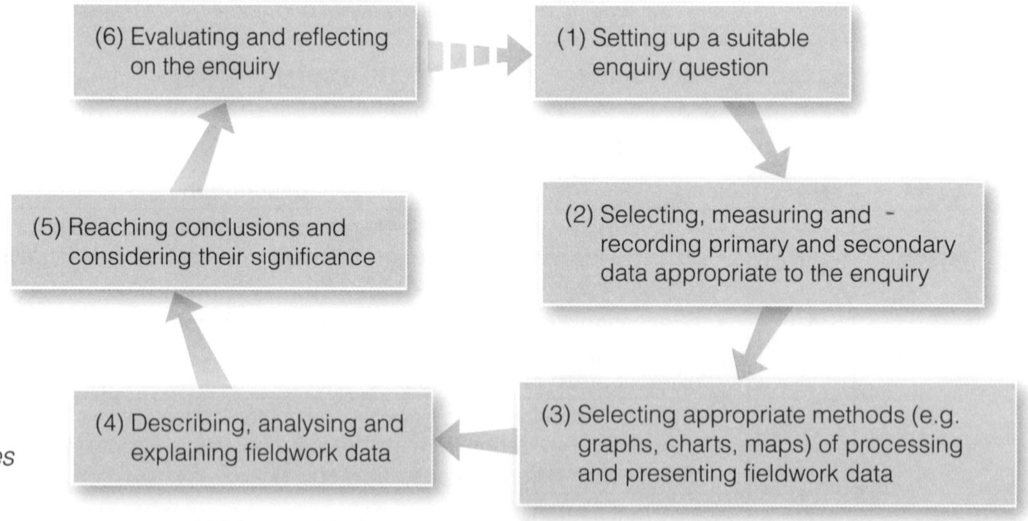

Figure 1 *The six stages of fieldwork enquiry*

(6) Evaluating and reflecting on the enquiry

(1) Setting up a suitable enquiry question

(2) Selecting, measuring and recording primary and secondary data appropriate to the enquiry

(3) Selecting appropriate methods (e.g. graphs, charts, maps) of processing and presenting fieldwork data

(4) Describing, analysing and explaining fieldwork data

(5) Reaching conclusions and considering their significance

Checklist!

The topic we studied for our **physical** fieldwork was _____

The place we went to was _____

The data we collected included _____

The topic we studied for our **human** fieldwork was _____

The place we went to was _____

The data we collected included _____

You need to know:

- how to prepare for familiar fieldwork questions in Paper 3.

What sort of questions will I be asked?

Paper 3 will give you the chance to write about *familiar fieldwork* – fieldwork you have carried out yourself. You've probably written up your fieldwork and, if possible, you should read through it as you prepare for the exam.

Examiners can set questions on any of the six stages of fieldwork enquiry in 24.1. To answer these questions well, *don't* spend time learning off by heart what you did – instead think more about what you learnt.

In the exam, you'll find questions about *familiar fieldwork* – that is, about fieldwork that you've done. They won't be testing your memory or recall.

These questions will tend to ask about the geographical skills that you used, and will expect you to think critically about what you did (such as how your fieldwork could be improved).

Familiar fieldwork questions will expect you to know and understand the fieldwork that you've done. The sorts of questions that could be set are shown in Figure **1**, with different questions for different stages of a geographical enquiry.

Stage	Familiar fieldwork questions that could be asked
Designing a question for enquiry	Explain one advantage of the location used for your human fieldwork enquiry. **[2 marks]**
Selecting, measuring and recording data	Describe one data collection technique that you used in your physical geographical fieldwork. **[2 marks]**
Processing and presenting fieldwork data	Justify one method that you used to present your primary data in your physical geography enquiry. **[2 marks]**
Describing and analysing fieldwork data	Explain two ways in which you analysed your human fieldwork data. **[4 marks]**
Drawing conclusions from your fieldwork	Assess the validity of your conclusions as a result of your human geography investigation. **[6 marks]**
Evaluating your fieldwork enquiry	Assess the effectiveness of the methods used in your physical geography investigation. **[9 marks]**

Figure 1 *Possible questions for each enquiry stage of your familiar fieldwork*

 **Six Second Summary**

Make sure that you:
- have revised your two pieces of fieldwork
- can recognise exam questions based on familiar fieldwork
- can recognise questions in the exam based on the six stages of fieldwork enquiry

 **Over to you**

Check through each of your two fieldwork write-ups. It will help you to have the following headings:
- Introduction to the area
- Aims
- Methods we used
- Data presentation
- Data analysis
- Conclusions
- Evaluation

You need to know:

- how to prepare for unfamiliar fieldwork questions in Paper 3.

Student Book
See pages 54–5

What sort of questions will I be asked?

In the exam, you'll also find questions about *unfamiliar fieldwork* – that is, about fieldwork in other, unfamiliar locations. These questions will tend to ask about the ways in which you can apply what you've learnt to new locations and situations (such as how you could apply skills you have learnt to a new situation).

Unfamiliar fieldwork questions differ in style from questions for familiar fieldwork. Your knowledge about the place won't be tested but questions will focus on how you apply what you've learnt to this new place. Questions could ask you about physical

Figure 2 An urban area where human fieldwork could be carried out

or human geography – so be prepared for both! For example, a photo of a city centre, like the one of York in Figure **2**, could be given and you could be asked questions about carrying out fieldwork there.

To make sense of the photo, think about the different stages of a geographical enquiry in Figure **1** in 24.1. The sorts of questions that could be set are shown in Figure **3** below, with different styles of questions and command words for different stages of an enquiry.

Stage	Unfamiliar fieldwork questions that could be asked
Designing a question for enquiry	Suggest one geographical fieldwork question that could be investigated in the area shown in Figure **2**. **[1 mark]**
Selecting, measuring and recording data	Describe one data collection technique that could be used in a geographical fieldwork investigation in the area shown in Figure **2**. **[2 marks]**
Processing and presenting fieldwork data	Calculate a) the mean, and b) the median values for the data shown. **[2 marks]**
Describing and analysing fieldwork data	Describe the pattern of pebble size shown on the completed graph. **[4 marks]**
Drawing conclusions from your fieldwork	To what extent can the following conclusion be drawn from the data: 'Environmental quality improves with distance from the city centre'? **[6 marks]**
Evaluating your fieldwork enquiry	Suggest one reason why the data shown in the graph may not be accurate. **[2 marks]**

Figure 3 *Possible questions for each enquiry stage of your unfamiliar fieldwork*

 **Six Second Summary**

Make sure that you:
- have revised your **two** pieces of fieldwork
- can recognise exam questions based on familiar fieldwork
- can recognise questions in the exam based on the six stages of fieldwork enquiry.

 **Over to you**

Check through each of your two fieldwork write-ups. It will help you to have the following headings:
- Introduction to the area
- Aims
- Methods we used
- Data presentation
- Data analysis
- Conclusions
- Evaluation

Student Book
See pages 307 and 317

You need to know:

- identify suitable questions for an enquiry
- say why a particular place would be suitable for fieldwork
- identify risks in a fieldwork environment, and how risks can be reduced.

Developing enquiry questions

Look at Figure **1**, a photo of Walton-on-the-Naze in Essex. You might not know it – just as the exam could include questions about places you've never been to. Remember that it's your ability to look at places and think like a geographer that is assessed in the exam.

A good fieldwork enquiry depends upon having a good question. It will probably be linked to a topic, for example, landscapes (see chapters 10, 11 and 12) or urban geography in the UK (see chapter 14).

- An example for coastal landscapes (Figure **1**) could be: *'How successful is coastal management along the coast at Walton-on-the-Naze?'*

Figure 1 *Coastal management at Walton-on-the-Naze in Essex*

Why are some places suitable for fieldwork?

You need to be able to say why a place may be suitable for fieldwork. Look at Figure **1**. What makes Walton-on-the-Naze good for fieldwork? You could say that:

- it has many different types of coastal management
- it is possible to see how well these types of coastal management protect the coast.

Now think about your own fieldwork. Identify features about **one** place you went to, which show why it was suitable for your investigation question. Write its name and your answers in the side panel.

Checklist!

My investigation question was _____

_____

The place we went to was _____

_____

What makes my place good for investigating my question? _____

Two potential risks we found were:

1 _____

_____

2 _____

_____

Identifying risks

You need to be able to say why a place is risky for fieldwork. In Figure **1**, there are risks from waves at high tide, cold weather or slipping on the rocks. Consider how these might be reduced by, for example, avoiding visiting when tides are high.

Now think about two risks in the place you visited. Write these in the Checklist.

 Six Second Summary

- Good fieldwork means having good enquiry questions in places which make them suitable for fieldwork.
- Some places are risky for fieldwork.

Over to you

Make a copy of the Checklist and complete it for the second fieldwork location you went to.

Student Book
See pages 308–9
and 318–9

You need to know:

- the differences between a) primary and secondary data and b) quantitative and qualitative data
- what data you collected for your fieldwork, and why
- different sampling methods
- how to describe and justify your methods of data collection.

Collecting fieldwork data

The point of fieldwork is to collect your own data. There are two types:

- *Primary data* are those you collect first-hand. These help you learn field techniques, for example, the speed of a river.
- *Secondary data* are those published by someone else. These help to balance the data you have collected yourself.

Each needs to be justified, so you should give reasons why you collected the data.

Primary and secondary data can be one of two categories:

- *Quantitative data* are 'hard' (or objective) data – including statistics. They come from making measurements (e.g. pebble size, or number of pedestrians).
- *Qualitative data* are 'soft' (or subjective) data which come from asking people's opinions, taking photos or making sketches.

Sample size is important to each of these categories because it helps to know whether data you collected were representative. Three methods can be used:

- *Random* – means samples are chosen without any pattern; every pebble (or person) has an equal chance of being selected.
- *Systematic* – means having a system to collect data; for example, every 10 cm across a river or every 10th person.
- *Stratified* – means a sample made up of different parts; for example, selecting different pebble sizes from a river so that you include the whole range of sizes found there. The same goes for gender or age groups in a town.

Checklist!

1. One example of primary *quantitative* data I collected was

   _____

   I collected this to help me find out _____

2. One example of primary *qualitative* data I collected was

   _____

   I collected this to help me find out_____

3. One example of secondary data I collected was

   _____

   I collected this to help me find out_____

4. A sampling method I used was _____

   which I used to collect data on _____

 Six Second Summary

- The purpose of fieldwork is to collect data suitable for a study.
- Primary data are first-hand; secondary data come from other sources.
- Quantitative data are objective (including statistics), qualitative data are subjective.
- Different sampling methods help to make fieldwork more accurate.

 Over to you

Copy and complete the Checklist for each of your fieldwork investigations.

Student Book
See pages 310–11
and 320–1

You need to know:

- about a range of visual, graphical and cartographic methods of presentation
- how to select and use methods of presentation accurately
- how to describe, explain and adapt different methods of presenting data.

Which method to use?

There are so many methods of presenting your fieldwork data visual.
Figure **1** summarises four types:

- Cartographic (maps). These show distributions.
- Visual, such as photos or sketches. Images help to illustrate and explain fieldwork.
- Tables of data. Useful when you are a member of a group and need to collect all your data together (this is called *collating* the data).
- Graphical, which can illustrate categories or change over time.

You should be clear about *which* techniques you chose to present your data, and *why*.

- *Continuous data* show change along a line of study (e.g. velocity along a stream or pebbles on a beach – a line graph may be best).
- *Categories* show classifications (e.g. pebble sizes grouped into sizes – a bar chart would be best here).
- If your sample sizes are different (e.g. 15 pebbles at one place, 17 at another), turn numbers into percentages and show using a pie chart.
- Make your data geographical by locating them on a map or aerial photo.
- *Annotated photographs* show evidence of processes (e.g. beach sand collecting near a groyne).
- *Field sketches* can show positive or negative aspects (e.g. an urban area).

Method	When you'd use it	Advantages
Maps / Cartography	Shows locations and patterns.	It is easier to compare patterns at locations.
GIS and photographs	Shows change over time (e.g. coastal erosion, or changes to a town).	Used to map data (e.g. census data) or aerial photos.
Table(s) of data	Can show raw data that you and your group collected.	Can help to identify anomalies (any data which look unusual).
Graphs and charts	When a picture tells a story better than a table (e.g. comparing two places).	Show data and patterns clearly – easier to read and compare than a table.

Figure 1 *When to use different methods of presenting data*

Checklist!

1a One method I used to present my primary *quantitative* data was

_____

I chose this method because

_____

1b An alternative method I could have

used was _____

which might have been better

because _____

2a One method I used to present my primary *qualitative* data was

_____

I chose this method because

_____

2b An alternative method I could have

used was _____

which might have been better

because _____

Six Second Summary

- There's a range of visual, graphical and cartographic methods of presentation.
- It's important to know why you chose different methods of presentation.

Over to you

Copy and complete the Checklist for each of your fieldwork investigations.

Student Book
See pages 312–13
and 322–3

You need to know:

- how to describe, analyse and explain your data
- how to identify anomalies
- how to use suitable statistical techniques.

What does analysis mean?

Analysing means:

- describing what your data show (e.g. highest and lowest points on a graph, or highest or lowest categories on a map)
- suggesting explanations (e.g. why pebble size might change along a beach, or why there are more pedestrians in place X than in place Y)
- making links between different sets of data (e.g. why pebble size and roundness change at the same time)
- identifying *anomalies* – that is, unusual data which don't fit the general pattern of results and suggesting why this might be so.

Using statistical techniques

Quantitative techniques, such as the mean (average) are used to analyse data. These include using a dispersion diagram to find the following:

- *range* – the difference between highest and lowest values
- *mode* – the number (or category) that appears most frequently in a data set
- *median* – to find this, place the data in rank order on a scaled line, and find the middle value. This divides the data into two equal halves.
- *quartiles* divide the data into four equal groups (two above and two below the median).

Using qualitative techniques

Sketches or photographs provide evidence about the place where you carried out your fieldwork. So they should be used to analyse your findings. Use the following:

- annotations to a photograph or sketch to highlight the main features of a place (Figure **1**)
- speech bubbles to illustrate people's quotes or opinions about an issue.

Checklist!

1a One method I used to analyse my *quantitative* data was

_____

I chose this method because

_____

1b An alternative method I could have

used was _____

which might have been better

because _____

2a One method I used to present my *qualitative* data was

_____

I chose this method because

_____

2b An alternative method I could have

used was _____

which might have been better

because _____

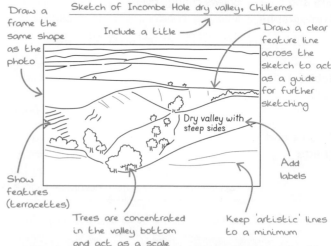

Figure 1 *Annotating a sketch of a photograph*

Draw a frame the same shape as the photo

Sketch of Incombe Hole dry valley, Chilterns

Include a title

Draw a clear feature line across the sketch to act as a guide for further sketching

Dry valley with steep sides

Add labels

Show features (terracettes)

Trees are concentrated in the valley bottom and act as a scale

Keep 'artistic' lines to a minimum

Six Second Summary

Analysis means:
- describing, analysing and explaining your data
- identifying anomalies
- use suitable statistical techniques.

Over to you

Copy and complete the Checklist for each of your fieldwork investigations.

Student Book
See pages 312–13
and 322–3

You need to know:

- how to draw conclusions based on fieldwork evidence
- how to link your conclusions to your aims.

What makes a good conclusion?

A conclusion is almost the end point of your fieldwork investigation. A conclusion is shorter than the analysis, because it is focused on the main enquiry question.

To write a proper conclusion, go back to your enquiry question (see 24.4).
Try to answer the following questions about each of your physical and human enquiries.

- What have you found out? Can you give an answer to the question?
- Which primary data did you collect that support your conclusion most strongly?
- Did any secondary data confirm that your conclusion might be right?
- Would any other evidence help to make it stronger?

Anomalies get in the way!

You'll probably have some results that don't quite fit the pattern of the rest. These are called *anomalies*. You need to comment on these and any other unexpected results.

- What might have caused these anomalies?
- Do they make your conclusion less certain? If so, why?
- Can they be ignored? If so, why?

Looking beyond

The final part of a conclusion is about the wider geographical significance of your study.

- Why might your study be important?
- Would any of your results be useful to other people or organisations such as local businesses or councils?
- Have you found out something that would apply to other places. For example, would all rivers be like yours, or all town centres show similar results?

Checklist!

1 One conclusion I can draw from my **physical** fieldwork is _____

2 This conclusion is firm/tentative (*cross out one*) because_____

3 The data which support my conclusion most are _____

4 The anomaly which makes my conclusion less certain is _____

_____

5 My conclusion would be even firmer if I could find out _____

_____

_____

Six Second Summary

A good conclusion means that you need:
- evidence to support it
- to consider any anomalies
- to think what other evidence could make it stronger.

Over to you

Copy and complete the Checklist for **each** of your fieldwork investigations.

*Student Book
See pages 314–15
and 324–5*

You need to know:

- how to identify problems and limitations of any methods you used
- how to identify any limitations of your data
- whether your conclusions are reliable.

What does 'evaluation' mean?

The evaluation is the last part of your fieldwork investigation, and it's probably the hardest. Remember that no study is perfect. You need to look back over the investigation and think about:

- how and when you collected data
- how the methods you used might have affected your results
- whether your conclusion is reliable – if you did it again in the same way, on a different day, would you get similar results?

Figure **1** should give you some ideas.

Checklist!

1 A possible source of error in my physical investigation was

_____

2 The ways in which these errors might have affected our results are

_____

3 My conclusions might be affected because

_____

Possible sources of error	Impacts on quality	Tick if this applies
Sample size	Were your sampling sizes large enough? • *Smaller sample sizes usually mean lower quality data.*	
Frequency of sample	Did you have enough sampling points (e.g. every 10m instead of every 100m)? • *Fewer sites reduce frequency and quality.*	
The type of sampling used (see 24.5)	Did you use random, systematic or stratified samples? • *The method you chose might create 'gaps' and introduce bias (e.g. if you only questioned people aged over 65).*	
Using the right equipment	The wrong/inaccurate equipment can affect overall quality by producing incorrect results. • *Is a dog biscuit the best way to record river velocity?* • *Did your questionnaire ask questions that gave you the answers you needed?*	
Time of survey	Times affect results (e.g. tides might influence beach accessibility and its measurable width). • *11.30 a.m. on a Monday in January may not give a sample of the whole population.*	
Location of survey	Big variations in beach profiles and sediment can occur in locations close to each other. • *Where you collect data matters.*	
Quality of secondary data	Did you find up-to-date secondary sources? • *Age and reliability of secondary data affect its quality.*	

Figure 1 *Possible sources of error in a geographical enquiry*

 **Six Second Summary**

Evaluation means looking critically at:
- how and when you collected data
- how methods might have affected results
- whether conclusions are reliable.

 Over to you

Copy and complete the Checklist for **each** of your fieldwork investigations.

You need to know:

- how to prepare for a range of question styles in Paper 3 Section B.

Getting it right

Fieldwork is worth getting right – it's worth 15% of the total mark for your GCSE.

To do yourself justice, remember a few simple points.

- You've done two days fieldwork – one physical, one human.
- There will be two sets of fieldwork questions – one on familiar fieldwork and the other on unfamiliar fieldwork. Physical and human fieldwork will be woven into these.
- No questions will ask you to describe your fieldwork – every question will be about your fieldwork *skills*, or will ask you to *apply* what you have done.

Now you're ready to try the questions below! Good luck!

 **Over to you**

A Familiar fieldwork

1 Explain why the location used for your physical fieldwork was suitable. **[2 marks]**

2 Explain **one** reason why **one** method of data collection was chosen for your human fieldwork. **[2 marks]**

3 Explain why you chose **one** particular method to present the results of your physical fieldwork. **[2 marks]**

4 Explain how one statistical method helped you to analyse **either** your physical **or** your human fieldwork data. **[4 marks]**

5 Assess how far your results enabled to you draw sound conclusions from your physical fieldwork. **[6 marks]**

6 Assess the reliability of methods you used to collect data in **either** your physical **or** your human fieldwork. **[9 marks]**

B Unfamiliar fieldwork

1 Explain one reason why the location for your physical fieldwork was suitable. **[2 marks]**

2 Explain how **one** potential risk presented by **one** location for your fieldwork was overcome. **[2 marks]**

3 Justify a method used to present **either** one quantitative **or** one qualitative set of data collected during your fieldwork. **[4 marks]**

4 Describe how different graphs helped you to analyse data from **one** of your fieldwork investigations. **[4 marks]**

5 Discuss which sets of data were most helpful to you in drawing conclusions from **one** of your fieldwork investigation. **[6 marks]**

6 Discuss how far your conclusions from any **one** of your fieldwork investigations could be described as 'reliable'. **[9 marks]**

Glossary

Abrasion (1) Rocks carried along a river wear down the river bed and banks; (2) the sandpaper effect of glacial ice scouring a valley floor and sides

Adaptation Actions taken to adjust to natural events such as climate change, to reduce damage, limit the impacts, take advantage of opportunities, or cope with the consequences

Aeroponics Growing plants in an air or mist environment without the use of soil

Appropriate (or intermediate) technology Technology suited to the needs, skills, knowledge and wealth of local people and their environment

Arch A wave-eroded passage through a small headland. This begins as a cave which is gradually widened and deepened until it cuts through

Arête A sharp, knife-like ridge formed between two corries cutting back by processes of erosion and freeze thaw

Attrition Rocks being carried by the river smash together and break into smaller, smoother and rounder particles

Bar Where a spit grows across a bay, a bay bar can eventually enclose the bay to create a lagoon

Beach A zone of deposited material that extends from the low water line to the limit of storm waves

Beach nourishment Adding new material to a beach artificially, through the dumping of large amounts of sand or shingle

Beach re-profiling Changing the profile or shape of the beach

Biodiversity The variety of life in the world or a particular ecosystem

Biomass Renewable organic materials, such as wood, agricultural crops or wastes, especially when used as a source of fuel or energy

Biotechnology The genetic engineering of living organisms to produce useful commercial products

Birth rate The number of births a year per 1000 of the total population

Brownfield site Land that has been used, abandoned and now awaits reuse; often found in urban areas

Bulldozing The pushing of deposited sediment by the snout (front) of the glacier as it advances

Business park An area of land occupied by a number of businesses

Carbon footprint Measurement of the greenhouse gases individuals produce, through burning fossil fuels

Cave A large hole in a cliff caused by waves forcing their way into cracks in the cliff face

Channel straightening Removing meanders from a river to make it straighter

Chemical weathering The decomposition (or rotting) of rock caused by a chemical change within that rock

Cliff a steep high rock face formed by weathering and erosion

Climate change A long-term change in the earth's climate, especially a change due to an increase in the average atmospheric temperature

Commercial farming Growing crops or raising livestock for profit, often involving vast areas of land

Commonwealth The Commonwealth is a voluntary association of 53 independent and equal sovereign states, most being former British colonies

Conservation Managing the environment in order to preserve, protect or restore it

Conservative plate margin Two plates sliding alongside each other, in the same or different directions — sometimes known as a transform plate margin

Constructive plate margin Tectonic plate margin where rising magma adds new material to plates that are diverging or moving apart

Consumer Organism that eats herbivores and/or plant matter

Corrie or cirque Armchair-shaped hollow in the mountainside formed by glacial erosion, rotational slip and freeze-thaw weathering — this is where the valley glacier begins

Cross profile The side-by-side cross section of a river channel and/or valley

Dam and reservoir A barrier built across a valley to interrupt river flow and create a man-made lake to store water and control river discharge

Death rate The number of deaths in a year per 1000 of the total population

Debt crisis When a country cannot pay its debts, often leading to calls to other countries for assistance

Debt relief Cancellation of debts to a country by a global organisation such as the World Bank

Decomposer Organisms such as bacteria or fungi that break down plant and animal material

Deforestation The cutting down and removal of forest

De-industrialisation The decline of a country's traditional manufacturing industry due to exhaustion of raw materials, loss of markets and overseas competition

Deposition Occurs when material being transported by the sea is dropped due to the sea losing energy

Dereliction Abandoned buildings and wasteland

Desertification The process by which land becomes drier and degraded, as a result of climate change or human activities, or both

Destructive plate margin Tectonic plate margin where two plates are converging and oceanic plate is subducted — there could be violent earthquakes and explosive volcanoes

Development The progress of a country in terms of economic growth, the use of technology and human welfare

Development gap Difference in standards of living and wellbeing between the world's richest and poorest countries

Discharge Quantity of water that passes a given point on a stream or river-bank within a given period of time

Drumlin Egg-shaped hill of moraine material deposited in a glacial trough

Glossary

Dune regeneration Building up dunes and increasing vegetation to prevent excessive coastal retreat

Earthquake A sudden or violent movement within the Earth's crust followed by a series of shocks

Economic impact Effect of an event on the wealth of an area or community

Economic opportunities Chances for people to improve their standard of living through employment

Ecosystem A community of plants and animals that interact with each other and their physical environment

Ecotourism Nature tourism usually involving small groups with minimal impact on the environment

Embankments Artificially raised river banks often using concrete walls

Energy conservation Reducing energy consumption by using less energy and existing sources more efficiently

Energy exploitation Developing and using energy resources to the greatest possible advantage, usually for profit

Energy mix Range of energy sources of a region or country, both renewable and non-renewable

Energy security Uninterrupted availability of energy sources at an affordable price

Environmental impact Effect of an event on the landscape and ecology of the surrounding area

Erratics Rocks transported and dumped by glacial ice to a different location, often hundreds of kilometres away

Erosion Wearing away and removal of material by a moving force, such as a breaking wave

Estuary Tidal mouth of a river where it meets the sea – wide banks of deposited mud are exposed at low tide

European Union A politico-economic union of 28 European countries – the UK is a member state

Extreme weather When a weather event is significantly different from the average or usual weather pattern, and is especially severe or unseasonal

Fair trade Producers in LICs given a better price for their goods such as cocoa, coffee and cotton

Famine Widespread, serious, often fatal shortage of food

Flood Where river discharge exceeds river channel capacity and water spills onto the floodplain

Floodplain Relatively flat area forming the valley floor either side of a river channel that is sometimes flooded

Floodplain zoning Identifying how a floodplain can be developed for human uses

Flood relief channels Artificial channels that are used when a river is close to maximum discharge; they take the pressure off the main channels when floods are likely

Flood warning Providing reliable advance information about possible flooding

Fluvial processes Processes relating to deposition, erosion, and transport by a river

Food chain Connections between different organisms (plants and animals) that rely upon one another as their source of food

Food insecurity Being without reliable access to enough affordable, nutritious food

Food miles The distance covered supplying food to consumers

Food security Access to sufficient, safe, nutritious food to maintain a healthy and active life

Food web A complex hierarchy of plants and animals relying on each other for food

Formal economy the type of employment where people work to receive a regular wage, pay tax, and have certain rights, i.e. paid holidays, sickness leave

Fossil fuel A natural fuel such as coal or gas, formed in the geological past from the remains of living organisms

Fragile environment An environment that is both easily disturbed and difficult to restore

Freeze-thaw weathering (or frost shattering) A common process of weathering in a glacial environment involving repeated cycles of freezing and thawing that can make cracks in rock bigger

Gabion Steel wire mesh filled with boulders used in coastal defences

Glacial trough Wide, steep-sided valley eroded by a glacier

Geothermal energy Energy generated by heat stored deep in the Earth

Globalisation Process creating a more connected world, with increases in the global movements of goods (trade) and people (migration & tourism)

Gorge A narrow steep-sided valley – often formed as a waterfall retreats upstream

Greenfield site A plot of land, often in a rural or on the edge of an urban area that has not been built on before

Green revolution An increase in crop production, especially in poorer countries, using high-yielding varieties, artificial fertilisers and pesticides

Grey water Recycled domestic waste water

Gross national income (GNI) Measurement of economic activity calculated by dividing the gross (total) national income by the size of the population

Groundwater management Regulation and control of water levels, pollution, ownership and use of groundwater

Groyne A wooden barrier built out into the sea to stop the longshore drift of sand and shingle, and allow the beach to grow

Hanging valley A tributary glacial trough on the side of a main valley often with a waterfall

Hard engineering Using concrete or large artificial structures to defend against natural processes, either coastal, fluvial or glacial

Hazard risk Probability or chance that a natural hazard may take place

Glossary

Headlands and bays A rocky coastal promontory (highpoint of land) made of rock that is resistant to erosion: headlands lie between bays of less resistant rock where the land has been eroded by the sea

High income country (HIC) A country with GNI per capita higher than $12 746 (World Bank, 2013)

Hot desert Parts of the world that have high average temperatures and very low precipitation

Human Development Index (HDI) A method of measuring development where GDP per capita, life expectancy and adult literacy are combined to give an overview

Hydraulic action Power of the water eroding the bed and banks of a river

Hydraulic power Process where breaking waves compress pockets of air in cracks in a cliff; the pressure may cause the crack to widen, breaking off rock

Hydroelectric power (HEP) Electricity generated by turbines that are driven by moving water

Hydrograph A graph which shows the discharge of a river, related to rainfall, over a period of time

Hydroponics Growing plants in water using nutrient solutions, without soil

Immediate responses Reaction of people as the disaster happens and in the immediate aftermath

Industrial structure Relative proportion of the workforce employed in different sectors of the economy

Inequalities Differences between poverty and wealth, as well as wellbeing and access to jobs, housing, education, etc.

Infant mortality Number of babies that die under one year of age, per 1000 live births

Informal economy employment outside the official knowledge of the government

Information technologies Computer, internet, mobile phone and satellite technologies

Infrastructure The basic equipment and structures (such as roads, utilities, water supply and sewage) that are needed for a country or region to function properly

Integrated transport system Different forms of transport are linked together to make it easy to transfer from one to another

Interlocking spurs Outcrops of land along the river course in a valley

Intermediate (or appropriate) technology Simple, easily learned and maintained technology used in LICs for a range of economic activities

International aid Money, goods and services given by single governments or an organisation like the World Bank or IMF to help the quality of life and economy of another country

Irrigation Artificial application of water to the land or soil

Landscape An extensive area of land regarded as being visually and physically distinct

Land use conflicts Disagreements between interest groups who do not agree on how land should be used

Lateral erosion Erosion of river banks rather than the bed – helps to form the floodplain

Levee Raised bank found on either side of a river, formed naturally by regular flooding or built up by people to protect the area against flooding

Life expectancy The average number of years a person is expected to live

Literacy rate Percentage of people in a country who have basic reading and writing skills

Local food sourcing Food production and distribution that is local, rather than national and/or international

Logging The business of cutting down trees and transporting the logs to sawmills

Long profile The gradient of a river, from its source to its mouth

Longshore drift Transport of sediment along a stretch of coastline caused by waves approaching the beach at an angle

Long-term responses Later reactions that occur in the weeks, months and years after the event

Low income country (LIC) A country with GNI per capita lower than $1045 (World Bank, 2013)

Managed retreat Controlled retreat of the coastline, often allowing flooding to occur over low-lying land

Management strategies Techniques of controlling, responding to, or dealing with an event

Mass movement Downhill movement of weathered material under the force of gravity

Meander A wide bend in a river

Mechanical weathering Physical disintegration or break up of exposed rock without any change in its chemical composition, i.e. freeze–thaw

Megacity An urban area with a total population of more than ten million people

Microfinance loans Very small loans which are given to people in the LICs to help them start a small business

Migration When people move from one area to another; in many LICS people move from rural to urban areas (rural–urban migration)

Mineral extraction Removal of solid mineral resources from the earth

Mitigation Action taken to reduce the long-term risk from natural hazards, such as earthquake-proof buildings or international agreements to reduce greenhouse gas emissions.

Monitoring (1) Recording physical changes, i.e. tracking a tropical storm by satellite, to help forecast when and where a natural hazard might strike; (2) using scientific methods to study coastal processes to help inform management options.

Moraine Frost-shattered rock debris and material eroded from the valley floor and sides, transported and deposited by glaciers

Natural increase Birth rate minus the death rate of a population

Glossary

Newly emerging economies (NEE) Countries that have begun to experience high rates of economic development, usually along with rapid industrialisation

North-south divide (UK) Economic and cultural differences between southern England and northern England

Nuclear power Energy released by a nuclear reaction, especially by fission or fusion

Nutrient cycling On-going recycling of nutrients between living organisms and their environment

Orbital change Changes in the pathway of the Earth around the Sun

Organic produce Food produced without the use of chemicals such as fertilisers and pesticides

Outwash Sediment deposited by meltwater that is well sorted and rounded in front of a glacier

Over abstraction When water is used more quickly than it is being replaced

Over-cultivation Where the intensive growing of crops exhausts the soil leaving it barren

Overgrazing Feeding too many livestock for too long on the land, so it is unable to recover its vegetation

Oxbow lake An arc-shaped lake on a floodplain formed by a cut-off meander

Permafrost Permanently frozen ground, found in polar and tundra regions

Planning Actions taken to enable communities to respond to, and recover from, natural disasters

Plate margin The border between two tectonic plates

Plucking A process of erosion – rocks are pulled from the valley floor as water freezes them to a glacier

Polar The most extreme cold environment with permanent ice, i.e. Greenland and Antarctica

Pollution Chemicals, noise, dirt or other substances which have harmful or poisonous effects on an environment

Post-industrial economy The shift of some HIC economies from producing goods to providing services

Precipitation Moisture falling from the atmosphere – rain, sleet or snow

Primary effects Initial impact of a natural event on people and property, caused directly by it, i.e. the buildings collapsing following an earthquake

Producer An organism or plant that is able to absorb energy from the sun through photosynthesis

Protection Actions taken before a hazard strikes to reduce its impact, such as educating people or improving building design

Pyramidal peak Where several corries cut back to meet at a central point, the mountain takes the form of a steep pyramid

Renewable energy sources A resource that cannot be exhausted, i.e. wind, solar and tidal energy

Resource management Control and monitoring of resources so that they do not become exhausted

Ribbon lake A long narrow lake in the bottom of a glacial trough

Rock armour Large boulders deliberately dumped on a beach as part of coastal defences

Rotational slip Slippage of ice along a curved surface

Rural-urban fringe A zone of transition between a built-up area and the countryside, where there is often competition for land use

Saltation Hopping movement of pebbles along a river or sea bed

Sand dune Coastal sand hill above the high tide mark, shaped by wind action

Sanitation Measures designed to protect public health, such as providing clean water and disposing of sewage and waste

Science park A collection of scientific and technical knowledge-based businesses located on a single site

Sea wall Concrete wall aiming to prevent erosion of the coast by reflecting wave energy

Secondary effects After-effects that occur as indirect impacts of a natural event, sometimes on a longer timescale, i.e. fires due to ruptured gas mains, resulting from the ground shaking

Selective logging Sustainable forestry management where only carefully selected trees are cut down

Service (tertiary) industries The economic activities that provide various services – commercial, professional, social, entertainment and personal

Shale gas Natural gas that is found trapped within shale formations of fine-grained sedimentary rock

Sliding Loose surface material becomes saturated and the extra weight causes the material to become unstable and move rapidly downhill

Slumping Rapid mass movement where a whole segment of a cliff moves down-slope along a saturated shear-plane or line of weakness

Social deprivation The extent an individual or an area lacks services, decent housing, adequate income and employment

Social impact The effect of an event on the lives of people or community

Social opportunities The chances available to improve quality of life, i.e. access to education, health care, etc.

Soft engineering Managing erosion by working with natural processes to help restore beaches and coastal ecosystems or to reduce the risk of river flooding

Soil erosion Removal of topsoil faster than it can be replaced, due to natural (water and wind action), animal, and human activity

Solar energy Sun's energy exploited by solar panels, collectors or cells to heat water or air or to generate electricity

Solution (or corrosion) Chemical erosion caused by the dissolving of rocks and minerals by river or sea water

Spit Depositional landform formed when a finger of sediment extends from the shore out to sea, often at a river mouth

Squatter settlement An area of (often illegal) poor-quality housing, lacking in services like water supply, sewerage and electricity

Glossary

Stack Isolated pillar of rock left when the top of an arch has collapsed

Subsistence farming A type of agriculture producing only enough food and materials for the benefit of a farmer and their family

Suspension Small particles carried in river flow or sea water, i.e. sands, silts and clays

Sustainability Actions that meet the needs of the present without reducing the ability of future generations to meet their needs

Sustainable energy supply Energy that can potentially be used well into the future without harming future generations

Sustainable food supply Food production that avoids damaging natural resources, providing good quality produce and social and economic benefits to local communities

Sustainable water supply Meeting the present-day need for safe, reliable and affordable water without reducing supply for future generations

Tectonic hazard Natural hazard caused by the movement of tectonic plates (i.e. volcanoes and earthquakes)

Till Sediment deposited by a glacier that is unsorted and angular

Traction Where material is rolled along a river bed or by waves

Trade Buying and selling of goods and services between countries

Traffic congestion When there is too great a volume of traffic for roads to cope with, and traffic slows to a crawl

Transnational corporation (TNC) A company that has operations (factories, offices, research and development, shops) in more than one country

Transportation The movement of eroded material

Tropical storm (hurricane, cyclone, typhoon) An area of low pressure with winds moving in a spiral around a calm central point called the eye of the storm – winds are powerful and rainfall is heavy

Truncated spur A former river valley spur which has been sliced off by a valley glacier, forming steep edges

Tundra A vast, flat, treeless Arctic region of Europe, Asia, and North America where the subsoil is permanently frozen

Undernutrition When people do not eat enough nutrients to cover their needs for energy and growth, or to maintain a healthy immune system

Urban farming Growing food and raising animals in towns and cities; processing and distributing food; collecting and re-using food waste

Urban greening Process of increasing and preserving open space in urban areas, i.e. public parks and gardens

Urbanisation When an increasing percentage of a country's population comes to live in towns and cities

Urban regeneration Reversing the urban decline by modernising or redeveloping, aiming to improve the local economy

Urban sprawl Unplanned growth of urban areas into the surrounding rural areas

Urban sustainability A city organised without over reliance on the surrounding rural areas and using renewable energy

Vertical erosion Downward erosion of the river bed

Volcano An opening in the Earth's crust from which lava, ash and gases erupt

Waste recycling Process of extracting and reusing useful substances found in waste

Waterfall A step in the long profile of a river usually formed when a river crosses over a hard (resistant) band of rock

Waterborne diseases Diseases like cholera and typhoid caused by micro-organisms in contaminated water

Water conflict Disputes between different regions or countries about the distribution and use of freshwater

Water deficit When demand for water is greater than supply

Water insecurity When water availability is insufficient to ensure the good health and livelihood of a population, due to short supply or poor quality

Water security Availability of a reliable source of acceptable quantity and quality of water

Water quality Measured in terms of the chemical, physical and biological content of the water

Water stress When the demand for water exceeds supply in a certain period or when poor quality restricts its use

Water surplus When water supply is greater than demand

Water transfer Matching supply with demand by moving water from an area with water surplus to another with water deficit

Wave cut platform Rocky, level shelf at or around sea level representing the base of old, retreated cliffs

Waves Ripples in the sea caused by the transfer of energy from the wind blowing over the surface of the sea

Wilderness area A natural environment that has not been significantly modified by human activity

Wind energy Electrical energy produced from the power of the wind, using windmills or wind turbines

Command words

It's important to answer questions properly. When you first read a question, check out the command word – that is, the word that tells you what to do.

The following is a list of command words and their meanings that are relevant to GCSE 9-1 Geography AQA.

Command word	What is it asking you to do?	Here's an example…
Assess	Weigh up which is the most/least important.	Choose either an earthquake or a volcanic eruption. Assess the extent to which primary effects are more significant than secondary effects.
Calculate	Work out.	Using the data in Figure 9, calculate the interquartile range of the pebble size data.
Compare	Identify similarities and differences.	Using Figure 4, compare HDI values in Africa and South America.
Complete	Add information to finish the task.	Using Figure 10, complete the graph for Dartmoor using the following data for rainfall.
Describe	Say what something is like.	No explanation is needed. Describe the distribution of hot deserts shown in Figure 6.
Discuss	Give the points on both sides of an argument and come to a conclusion.	Discuss the effects of urban sprawl on people and the environment.
Evaluate	Make judgements about which is most or least effective.	Evaluate the effectiveness of an urban transport scheme you have studied.
Explain	Give reasons why something is the case.	Explain how food security can be improved.
Identify	Name an example, sometimes from a map, photo or graph.	Identify two data collection techniques that could be used to carry out a geographical fieldwork investigation in one of the areas shown.
Justify	Give evidence to support your ideas.	Do you agree with this statement? Justify your decision.
Outline	Summarise the main points.	Outline one reason why the concentration of carbon dioxide in the atmosphere has changed over time.
Suggest	Give a well-reasoned guess to explain something where you can't be sure of the answer.	Using Figures 11 and 12, suggest why there might be a need for water transfer from one part of the UK to another.
To what extent...?	Judge the importance of something.	To what extent do urban areas in LICs or NEEs provide social and economic opportunities for people?
Use evidence to support this statement	Choose information to prove or disprove something.	'Weather in the UK is becoming more extreme.' Use evidence to support this statement.

Please note that this our interpretation of AQA's guidance and is not an exhaustive list of all command words that might be used in exams

Symbols on Ordnance Survey maps (1:50 000 and 1:25 000)

ROADS AND PATHS

M I or A 6(M) — Motorway
A 35 — Dual carriageway
A 31(T) or A 35 — Trunk or main road
B 3074 — Secondary road
Narrow road with passing places
Road under construction
Road generally more than 4 m wide
Road generally less than 4 m wide
Other road, drive or track, fenced and unfenced
Gradient: steeper than 1 in 5; 1 in 7 to 1 in 5
Ferry; Ferry P – passenger only
Path

PUBLIC RIGHTS OF WAY

(Not applicable to Scotland)

1:25 000	1:50 000	
		Footpath
		Road used as a public footpath
+-+-+-+-+		Bridleway
	-+-+-+-+-	Byway open to all traffic

RAILWAYS

Multiple track
Single track
Narrow gauge/Light rapid transit system
Road over; road under; level crossing
Cutting; tunnel; embankment
Station, open to passengers; siding

BOUNDARIES

+ — + — + — National
+ + + + + District
— · — · — County, Unitary Authority, Metropolitan District or London Borough
National Park

HEIGHTS/ROCK FEATURES

50 — Contour lines
· 144 — Spot height to the nearest metre above sea level

outcrop   cliff   scree

ABBREVIATIONS

P	Post office	PC	Public convenience (rural areas)
PH	Public house	TH	Town Hall, Guildhall or equivalent
MS	Milestone	Sch	School
MP	Milepost	Coll	College
CH	Clubhouse	Mus	Museum
CG	Coastguard	Cemy	Cemetery
Fm	Farm		

ANTIQUITIES

Roman — ⚔ Battlefield (with date)
Non-Roman — ⁎ *Tumulus/Tumuli* (mound over burial place)

LAND FEATURES

ruin — Buildings
Public building
Bus or coach station
Place of Worship { with tower / with spire, minaret or dome / without such additions }
∘ Chimney or tower
Glass structure
Ⓗ Heliport
△ Triangulation pillar
Mast
Wind pump / wind generator
Windmill
Graticule intersection
Cutting, embankment
Quarry
Spoil heap, refuse tip or dump
Coniferous wood
Non-coniferous wood
Mixed wood
Orchard
Park or ornamental ground
Forestry Commission access land
National Trust – always open
National Trust, limited access, observe local signs
National Trust for Scotland

WATER FEATURES

Marsh or salting   Slopes   Cliff   High water mark
Towpath Lock   Low water mark
Aqueduct   Canal   Flat rock   Lighthouse (in use)
Weir   Ford   Sand
Normal tidal limit   Dunes   Lighthouse (disused)   Beacon
Bridge   Mud   Shingle
Footbridge
Lake
Canal (dry)

TOURIST INFORMATION

Ⓟ Parking
P&R Park & Ride
Ⓥ Visitor centre
Information centre
Telephone
Camp site/ Caravan site
Golf course or links
Viewpoint
PC Public convenience
Picnic site
Pub/s
Museum
Castle/fort
Building of historic interest
Steam railway
English Heritage
Garden
Nature reserve
Water activities
Fishing
Other tourist feature
Moorings (free)
Electric boat charging point
Recreation/leisure/ sports centre

Revision planner

Date: _____

	Revision Period 1	Revision Period 2	Revision Period 3	Revision Period 4	Revision Period 5
Monday					
Tuesday					
Wednesday					
Thursday					
Friday					
Saturday					
Sunday					

Revision planner

Date: _____

	Revision Period 1	Revision Period 2	Revision Period 3	Revision Period 4	Revision Period 5
Sunday					
Saturday					
Friday					
Thursday					
Wednesday					
Tuesday					
Monday					